9 781016 011440

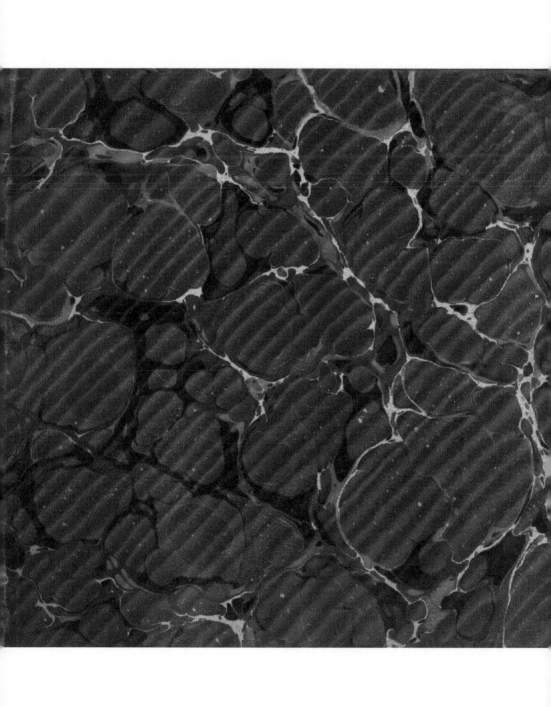

Sarica

# MILITARY SIGNS AND ILLUSTRATIONS OF MODERN FORTIF[ICATION]

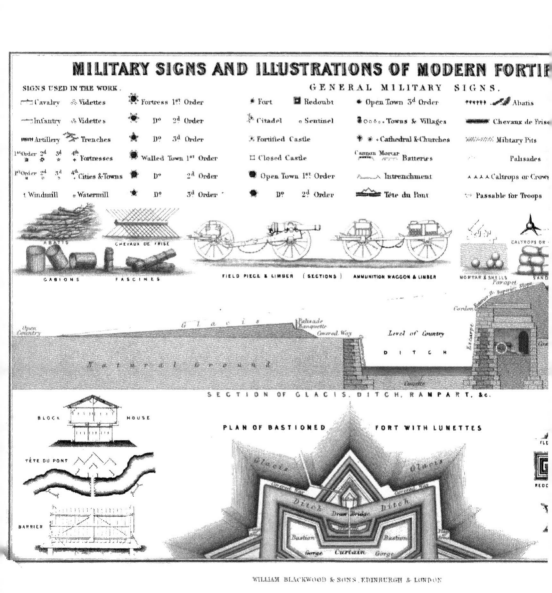

SIGNS USED IN THE WORK.

GENERAL MILITARY SIGNS.

Cavalry — Videttes — Fortress 1st Order — Fort — Redoubt — Open Town 3d Order — Abatis

Infantry — Videttes — Do 2d Order — Citadel — Sentinel — Towns & Villages — Chevaux de Frise

Artillery — Trenches — Do 3d Order — Fortified Castle — Cathedral & Churches — Military Pits

1st Order 2d 3d 4th Fortresses — Walled Town 1st Order — Closed Castle — Cannon Mortar Batteries — Palisades

1st Order 2d 3d 4th Cities & Towns — Do 2d Order — Open Town 1st Order — Intrenchment — Caltrops or Crows

Windmill — Watermill — Do 3d Order — Do 2d Order — Tête du Pont — Passable for Troops

ABATIS — CHEVAUX DE FRISE — CALTROPS OR

GABIONS — FASCINES — FIELD PIECE & LIMBER (SECTIONS) — AMMUNITION WAGGON & LIMBER — MORTAR & SHELLS — SAND

Open Country — Glacis — Palisade — Banquette — Covered Way — Level of Country — Parapet — Gordon — Escarp — DITCH — Cunette

Natural Ground

SECTION OF GLACIS, DITCH, RAMPART, &c.

BLOCK HOUSE

PLAN OF BASTIONED FORT WITH LUNETTES

FLE[CHE]

TÊTE DU PONT — Glacis — Glacis — RED[OUBT]

Ditch — Draw Bridge — Ditch

BARRIER — Bastion — Bastion

Gorge — Curtain — Gorge

WILLIAM BLACKWOOD & SONS, EDINBURGH & LONDON.

# ATLAS

TO

# ALISON'S HISTORY OF EUROP

CONSTRUCTED AND ARRANGED, UNDER THE DIRECTION OF

## MR ALISON,

BY

## ALEXANDER KEITH JOHNSTON, F.R.G.S.

AUTHOR OF THE NATIONAL, AND EDITOR OF THE PHYSICAL ATLAS.

WITH A

CONCISE VOCABULARY OF MILITARY AND MARINE TERMS.

WILLIAM BLACKWOOD AND SONS, EDINBURGH AND LONDO
MDCCCXLVIII.

PRINTED BY WILLIAM BLACKWOOD AND SONS, EDINBURGH.

# CONTENTS.

# CONTENTS.

A

# CONCISE VOCABULARY

OF

# MILITARY AND MARINE TERMS.

---

ABBATIS. A species of intrenchment made by trees cut down and laid in a ditch or other excavation, at a short distance from the parapets of field-works, with their branches pointing outwards, to prevent or retard the advance of an enemy.

ACCOUTREMENTS. This term denotes the belts, pouches, &c., of a soldier.

ADJUTANT. The Adjutant is the assistant of the commanding and field officers in the execution of their duty.

ADJUTANT-GENERAL OF THE FORCES. An officer of high rank and trust at the Horse Guards in London.

ADJUTANT-GENERAL. An officer of distinction, selected to assist the general of an army in his various duties.

ADVANCED GUARD. A detachment of troops preceding the march of the main body.

AIDE-DE-CAMP. An officer attached to a general, to carry orders, &c.

ALARM-POST. The place appointed for every detachment or regiment to assemble in case of a sudden alarm.

APPROACHES. The first, second, and third parallels, with their corresponding trenches, saps, and mines, by means of which the besiegers approach in comparative safety a fortified place.

APPUI, POINT D'. A term applied to any given point upon which a body of troops is formed.

ARMISTICE. A truce or temporary suspension of hostilities.

ARMOURER. A person who makes, repairs, and cl[e
ARMOURY. A storehouse in which arms are kept.
ARMS, A STAND OF. A complete set for one soldie[r
ARSENAL. A magazine for military stores.
BALLS. Spherical bodies made of cast-iron or lead
  *Cast-iron balls* are generally used by artillery,[ for musketry.
  *Light balls* are used at sieges, in order to dis[
    parties from the light given by them.

BANQUETTE. A small mound of earth, three or [f]
  elevated to within four or five feet of the crest [
  to enable the shortest men to fire over it with fi[

BARBET BATTERIES. Batteries without embrasure[s
  guns are raised to fire over the parapet, generall[y
  salient angles of the different works, to en[
  mounted on them to range freely over the adjac[

BARRACKS. Buildings for the lodgment of troops
  cessary accommodation for cooking, guard-room[
  magazines, &c.

BARRICADE. An obstruction formed in streets, a[
  as to block up access to an enemy. They are ge[
  of overturned waggons, carriages, large stone[s
  abbatis, &c.

BASTION. In fortification, generally constructed

angle of the polygon, forming the magistral or principal line of defence surrounding a place. It consists of two faces, right and left, and two corresponding flanks, and is so formed as to be well seen, and defended every where by the flanking fire of some other part of the works.—*Bastions* are distinguished into empty and full. The interior surface of the empty bastion is on a level with the *terreplein* or ground of the place. The interior of the full is raised by earth to the level of the rampart.

BASTIONED FORT. Generally a field-work constructed on a polygon, upon the principles of permanent fortification, of which the lines of defence do not exceed the range of musketry.

BATON. A short staff or truncheon borne by field-marshals as a symbol of their authority.

BATTALION. A body of infantry, generally composed of ten companies, each consisting of a lieutenant, an ensign, three or four sergeants, and about 100 rank and file, under the charge of a captain : the whole, with the staff-officers attached to it, such as adjutant, paymaster, quartermaster, surgeon, and assistants, being under the command of a lieutenant-colonel.

BATTERY. The name given to any place where cannon, mortars, &c., are mounted for the purpose of defending or attacking important points. Also, in the field, a division of a regiment or brigade of artillery, as a company is a division of a regiment of infantry.

BEAR UP OR AWAY. The act of changing a ship's course, so as to cause her to sail more before the wind than she did previously.

BEATING TO WINDWARD. Making a progress against the direction of the wind, by steering alternately close-hauled on the starboard and larboard tacks.

BERM. A narrow level space, two or three feet wide, along the exterior slope of a parapet, to prevent the mass of earth and other materials, of which it is composed, from falling into the ditch.

BIVOUAC. An army bivouacs at night when it does not encamp or take up quarters.

BLOCKADE. A place is said to be blockaded by land or sea when all ingress and egress is prevented by troops or ships of war surrounding it.

BLOCKHOUSE. Originally a work built nearly or wholly of the trunks of trees. Now applied generally to small forts, capable of protecting against musketry, but not against artillery.

BODY OF A PLACE. The space enclosed l formed by bastions, curtains, ravelins

BOMBARD. To throw bombs or shells i to its destruction, and to compel surr

BOOM. A strong beam of timber, &c., harbour to prevent the entrance of an

BREACH. An opening effected by artill and defences of a fortified place.

BREAK GROUND. To commence the sie trenches, &c.

BREASTWORK. A parapet thrown up to of the troops of a place, to protect the

BRIDGE. Besides the usual stone, cha there are several descriptions of mi boats connected together, pontoon bric according to emergencies. These are j &c., so as to give a safe passage for troo]

BRIGADE. A division of troops composed ing of detachments of infantry and cav of a general officer. The strength There are cavalry brigades and infan containing usually from six to nine s three to six battalions. There are a The term is also frequently applied, in troop of mules with their drivers, in t sariat or of store-keepers.

To BRING TO. To check the course of a cing, by arranging the sails in such counteract one another, and prevent or retreating.

BROADSIDE. A discharge of all the gu on both her upper and under decks.

BULKHEAD. A partition separating on other on the same deck.

CAISSON. A term used for various pur nifies a box or chest for holding amm at other times for musketry, having d ticular kinds. It is also used for de gons, &c.

CALIBRE or CALIBER. The diameter cannon or other firearm.

CAMP. The ground covered by an arm;

CANTEEN. A small circular tin or wooden vessel, used by soldiers to carry liquor when they are on active service.—A trunk or chest containing culinary and other utensils for the use of officers.—A suttling-house kept in garrisons for the use of the troops.

CANTONMENTS. When troops are detached and quartered in different adjacent towns and villages, they are said to be placed in cantonments.

CAPITAL. A line drawn from the angle of the polygon, forming the salient angle of the bastion to the middle of its gorge, &c.

CAPITULATION. The surrender of a fortress or army on stipulated conditions.

CAPONIERE. A protected passage from the body of the place to an outwork. It frequently forms a secure passage or covered way, made by a small glacis on each side from the middle of the curtain to the gorge of the ravelin. It also serves as a defence to the main ditch by a raking fire of musketry.

CARBINE. A short small musket, used principally by cavalry.

CARRIAGE of a gun. The machine upon which it is mounted.

CARRY. To obtain possession of any place by force, whether outworks, field-works, a battery, a parapet, or a town itself.

CARTRIDGE. A case of paper, flannel, or parchment, fitted to the bore of a piece, and containing an exact charge of gunpowder. These are blank-cartridges. The addition of the bullet in the same envelope constitutes the ball-cartridge.

CASE or CANISTER SHOT. Discharged from heavy ordnance, and consists of a number of musket bullets or iron balls enclosed in a tin or iron case.

CAVALIER. A work formed within a full bastion, and elevated ten or twelve feet above it, to command a particular point, and give additional strength to works.

CHAMBER. The place where the powder is deposited in a mine. The cavity, in a gun or mortar, which receives the charge.

CHARGE. In gunnery, comprehends the amount of powder and shot with which a gun is loaded.

CHEVAUX-DE-FRISE. An object employed in fortification for the defence of places. It consists of a prismatic beam of timber of a square or hexagonal form, of about six or eight feet long, and five or six inches in diameter, through which pointed stakes are driven perpendicular to each of its faces, equi-distant from each other, and radiating from the centre of the beam.

CHEVRONS. The bars or distinguishing marks on of non-commissioned officers.

CITADEL. A fortress, generally in the form of a pe gon, situated on the most commanding grou city, though generally separated by an espla ground without buildings, so that no approac unperceived.

CLOSE-HAULED. That trim of a ship's sails whicl make a progress in the nearest possible directi point whence the wind blows.

COLONEL. The first officer in command of a regim cavalry, or artillery.

COLOURS of a regiment. Two silk flags carried by th

COLUMN, in a military sense, is used in contradist Thus a regiment of cavalry is in line when its displayed. It may advance in column of squad (which are half-squadrons,) of divisions, (w troops,) or of threes, (according to the mode mation.) Troops moving along a road are column. Hence a body of troops on the m spoken of as "a column." "Close column squadrons, companies, or battalions, &c. &c. in rear of each other. "Open column" is interval is left for them to wheel into line if re

COMMAND. In the regular forces belongs to the Command, in fortification, is the elevation c work above the exterior, so as to see and fire or over the level country.

COMMUNICATION, LINES OF. Trenches made to intercourse between any two points or forts.

CONTRIBUTION. A tax paid to a hostile force, by of a town or country, to avoid being plundered

CONVOY. A guard of troops employed to escort p ammunition, or money, conveyed in time of wa or place to another. It is also a ship of war e tect a fleet of merchant ships during the whole voyage.

CORDON. A round projection of stone placed o revetement of the escarp, to throw the rain of and prevent the besiegers ascending by their escalade. A square projection, called a *tablette* preferred.

CORPORAL. Lowest grade of non-commissioned officers.

CORNET. Lowest rank of commissioned officers in cavalry regiments.

CORPS. This word, which has crept into our language from the French, means literally a body, but is variously applied. In common English parlance, it usually designates a regiment. In Napoleon's wars it was applied to large divisions of troops; to large armies, in fact, detached from still larger ones. In the Russian campaign some of the so-called "corps" numbered forty or fifty thousand men. By modern historians the word is generally used in the more extended sense.

COVER. In military operations, implies generally security or protection.

COVERED WAY. A space of about 30 feet broad, extending from the counterscarp of the ditch to the crest of the glacis, passing completely round the whole body and outworks of a place.

COUNTER-GUARD. A revêted work, consisting of a comparatively narrow rampart and parapet, commonly parallel to the faces of a bastion or ravelin, so as to strengthen any particular point liable to attack.

COUNTERSCARP. The outer boundary of the ditch, which is generally faced or revêted with masonry, to render the descent into the ditch difficult.

COUNTERSIGN. A watchword, demanded by sentries of those who approach their post.

COUP-DE-MAIN. A sudden and vigorous attack, for the purpose of instantaneously capturing a place or gaining a position.

CROWN-WORK. A figure resembling a crown, and consisting of two small fronts of fortification connected with the body of the place by two long sides, so as to occupy a position it is desirable to secure.

CUIRASS. A piece of metallic defensive armour, covering the more exposed and vital parts of the body, from the neck to the waist.

CUNETTE OR CUVETTE. A trench generally 7 or 8 feet deep, and 10 or 12 wide, serving to prevent the passage of troops through a dry ditch; whilst, at the same time, it carries off the superfluous water from the place.

CURTAIN. In fortification, is that portion of the rampart which connects two adjacent bastions.

DEBOUCH. To march out of a wood or defile into open ground.

DEFILE. A narrow passage, through which, in marching, troops can present a narrow front only, and therefore dangerous in presence of an enemy.

DEMILUNE. Called also a *ravelin*, is a ʋ the curtain and flanks of the bastion.

DESCENT. The landing of troops from t of invading a country.

DISLODGE. To drive an enemy from any ;

DITCH. In fortification, is an excavatic from which the earth required for the part and parapet is obtained.

DIVISION. A portion of an army, incl and artillery, and commanded by a g

DOUBLING. The act of sailing round or point of land.

DOUBLING UPON. In naval tactics, is the of a hostile fleet, and placing it betwe

ECHELON. This word is adopted from meaning in that language being a step mation, in the movements of an arm the steps of a ladder, and therefore g is very convenient for the attack and for oblique and direct changes of posit produced by the wheel of divisions thi of a circle.—*Direct changes* are prod and successive march of divisions fron

EMBRASURE. An opening through the p point a gun.

ENCEINTE. The rampart enclosing the consisting of bastions, curtains, and tl

ENFILADE. To sweep the whole length of ε by the fire of a battery formed on a p

ENSIGN. The lowest rank of infantry c ordinate to lieutenant. Ensigns carr ;

EPAULEMENT. An elevation of earth, thrown up perpendicularly to the fac€ troops behind it from an enfilade fire quently composed of fascines or gabio

EQUIPAGE, CAMP. Consists of tents, kitcl saddle-horses, baggage-waggons, &c.

ESCALADE. To attack a fortress by sca walls by means of ladders.

ESCARP. The side of the ditch next the manent fortification, is faced with st( revêtement.

ESPLANADE. A part of a fortified place set apart for exercise and public walk.

EVACUATE. To withdraw from a town or fort in consequence of treaty, capitulation, or orders.

EVOLUTION. A movement by which troops change their position for attack or defence.

FALSE ATTACK, OR FEINT. A pretended attack, made to deceive and divert an enemy from the real point to be assailed.

FASCINES. A species of long cylindrical faggots, made of brushwood or branches of trees, for the purpose of revêting the cheeks of embrasures, or supporting the earth of extensive epaulements in field-batteries.

FAUSSE BRAYE. A work formerly used in fortification, but now seldom employed, as its disadvantages counterbalance its advantages. It is a platform rising to half the height of the revêtement, thus giving a good grazing fire against the besiegers before they enter the ditch, but it then affords a ready means of scaling the walls.

FIELD-MARSHAL is the highest military rank in the British service.

FIELD OFFICERS. Colonels, lieutenant-colonels, and majors.

FILE. A line of soldiers drawn up behind one another. The general term means two soldiers, consisting of the front and rear rank men.

FIRELOCK. A general name for the infantry musket.

FLANK. The extreme right or left of a body of troops, or of an encampment.

FLANK COMPANIES. The right and left companies of infantry regiments, generally grenadiers and light infantry.

FLECHE or arrow. A simple species of field-work, consisting of two faces forming a salient angle, and easily constructed for the defence of a position.

FLEET. Any number of vessels above five sail of the line.

FORAGE. Provender for the horses of an army.

FORD. A shallow part of a river, where troops may pass without bridge or boat.

FORE AND AFT signifies throughout the whole ship's length.—To rake a ship fore and aft is to fire along her decks, thus causing the greatest possible damage and loss of life.

FORLORN HOPE. A party of officers and men who are detached to lead the attack when an army storms a fortress. From the great danger attending this operation its name is derived.

FORTS are works constructed to secure places of importance, to afford support to the wings or particular parts to command the resources of a district of country

FORTIFICATION. The art of enclosing towns or othe works, so as to render them more easily defensib rison. It may be divided into different classes, field, defensive, offensive, natural, and artificial.

FOSSE. The French name for a ditch.

FRAISES. Palisades or stakes ranged in an inclin rected towards the breasts of an enemy, and forr

FUSE. A tube fixed into a shell filled with con furnished with a quick match. It is made of portional to the distance intended to be thrown so as to burn during its flight, and to explode t moment it strikes the ground.

FUSIL. A small species of musket.

FUSILEERS. A body of troops originally arme which gave the name. There are now only for called in the British service, and which are no l

GABIONS. Cylindrical baskets of wicker-work, w and filled with earth. They form a conveni revêtement in field-works, especially during construction of batteries and the formation of s

GALLERY. A passage communicating to that pa which powder is placed.

GARRISON. The guard of a fortified place : th generally, the troops quartered in a town.

GENERAL OFFICERS. All those above the rank of

GENERALISSIMO. The commander-in-chief of a co

GENOUILLIERE. That part of the parapet in the er the platform and under the gun.

GLACIS. The superior slope of the parapet of th extended in a gentle slope of about 1 in 20 or of the surrounding country. Its length is al feet.

GORGE. The entrance from behind into a basti redoubt.

GRENADE. A small shell, whose fuse is set fire t then thrown by hand among the enemy, to a dis or thirty yards, causing considerable damage sion. It is now chiefly used against besiegers a breach.

GRENADIERS. The tallest and stoutest soldiers

of infantry, selected and formed into a company, posted on the right of the battalion, and leading it in attack.

GUARD. A certain portion of troops appointed to watch a position and prevent a surprise.

GUNNER. A soldier employed to manage and discharge great guns. An artilleryman. In the British artillery the private soldiers are divided into gunners and drivers.

GUNSHOT. Generally understood to be the point-blank range of a gun.

HAVERSACK. A coarse linen bag, issued to every soldier on service, for the purpose of carrying provisions.

HELMETS. Pieces of defensive armour for the head, chiefly worn by heavy cavalry.

HOLSTERS. Leathern cases fixed in the front of a saddle to contain a horseman's pistols.

HONOURS OF WAR. This expression is generally used in speaking of troops capitulating and evacuating a fortress. The nature of such " honours " depends very much on circumstances. In some cases the retreating forces depart scotfree, with colours, cannon, and baggage—in others, they retire to a distance, pile their arms, and then surrender as prisoners of war.

HORN WORK. Composed of two half bastions and a curtain, with two long sides directed perpendicularly upon the faces of the bastions or ravelins, so as to be defended by them. This disposition, however, has latterly been improved, by augmenting the number of fronts, and shortening the branches.

HOUSEHOLD TROOPS consist of the regiments of Life Guards, Horse Guards, and Foot Guards.

HOWITZER. A piece of ordnance of the nature of a mortar, of various calibres and dimensions.

INFANTRY. The foot-soldiers of an army.

INTRENCHMENT. A general term, denoting a ditch or trench, with a parapet for the purposes of attack or defence.

INVEST. To take the initiatory measures to besiege a town, by securing every road and avenue leading to it, and by seizing the commanding positions. The business of an investing force is to prevent the garrison from receiving assistance or supplies, and to retain its ground till the arrival of the army with its breaching artillery allows the commencement of the siege in due form.

INUNDATION. One of the most efficacious methods of impeding the approach of an enemy to any fortification or field-work.

It is effected by turning the course of situation covering the work, by means

KNAPSACK. A square frame covered wit pared for strapping on the infantry sol ing the whole of his regimental necess

KNOT. In navigation is a measure of about $\frac{1}{120}$ part of a nautical mile. minute glass is $\frac{1}{120}$ part of an hour. divided similarly to the hour, whateve off the reel in half a minute, by the sl water, the same number of nautical i hour. Knots and miles are therefo and are used synonymously.

LADDERS, SCALING. A particular kind staves or steps, for the purpose of scali the ramparts of an enemy.

LARBOARD. The left side of a ship w the head.

LARBOARD TACK. The situation of a shi wind blowing on her larboard, or left

LEE. That part of the hemisphere to wh

LEE-GAGE. A ship or fleet to leeward o the lee-gage.

LIE UNDER ARMS. To remain in a state :

LIEUTENANT. An officer in rank next ui

LIFE-GUARDS. Picked regiments suppos tended for the guard of the sovereign's precedence of every other corps in the

LIGHT INFANTRY. Regiments or compa strong men, and are generally distingu and efficient services in the field.

LIMBER, in artillery. The fore-part of a to which the horses are attached. WI the gun is unlimbered by unhooking t and the limber is taken away to a few

LINE. Troops of the line are all those army, excepting the Life Guards, Hor Royal Marines, Fencibles, Militia, V corps.

LINES. A species of field-works to cove tiers of a state, or a district round a t continued or broken with intervals.

shapes to produce a flanking fire for mutual defence and support, as is common in fortifications.

LITTER. A species of hurdle or palanquin-bed, in which those severely wounded are carried from the field of battle.

LOG LINE and LOG. Instruments by which the ship's velocity is measured.

LOG BOARD. That on which the daily transactions on board ship are recorded, whence they are copied into the log-book—the legal record of every nautical transaction.

LOOP-HOLES. Small openings, similar to embrasures in the walls of a citadel or fortification. Loop-holes are also made in the walls of gardens, or even houses, for the defence of important points during a battle, by a fire of musketry through them.

LUNETTES. Small works constructed to strengthen a ravelin or other part of a fortification. In this case, one face is about perpendicular to that of the ravelin, and the other, to that of the bastion, and so on in similar cases.

MAGAZINE, in general, is a place in which stores, arms, ammunition, and provisions, are kept. The name is frequently restricted to a place for preserving powder and shot.

MAJOR. An officer next in rank to the lieutenant-colonel of a regiment.

MAJOR-OF-BRIGADE. An officer, through whom orders are communicated to the troops, and considered as attached to the brigade, not to the officer commanding it.

MARINES. A body of troops especially for the naval service, trained to encounter an enemy either at sea or land.

MASK. A cover for a battery, so as to prevent it being seen and recognised by an enemy. When a body of troops encamps before a fortress, so as to prevent the garrison from moving out to harass an army acting freely in its vicinity, it is said to be masked by the hostile forces.

MATCH. A substance composed to retain fire for the service of artillery, mines, and fireworks. There are two kinds, quick and slow. Tow and sulphur are common ingredients.

MERLON. The space in the parapet between two embrasures, of about eighteen feet in length.

MINE. A subterraneous passage carried from the lines of the besiegers, under the rampart of a fortification, to blow it up by gunpowder.

MORTARS. Short cannon, of large bore, made of cast-iron or brass. They are used to throw shells, which, sion, set fire to buildings, overthrow works, dism destroy troops.

MUSKET-PROOF. Any object capable of resisting musket-balls is said to be musket-proof.

MUSTER. A review of troops under arms, fully equ to take an account of their numbers, inspect accoutrements, and examine their condition.

NATURAL FORTIFICATION, or STRENGTH. A combina obstructions, tending to impede the movements

NON-COMMISSIONED OFFICERS. The sergeant-major, sergeant, sergeants, corporals, and drum and fife appointed by order of the commanding officer o

OPENING of the trenches. The first act of breakin besieging army, for the purpose of carrying their to the place.

ORDNANCE. A name applied to every thing conn engineer and artillery service.—Cannon are fu nated pieces of ordnance.

OUTWORKS. All the works constructed beyond t place, such as ravelins, tenailles, covered ways,

PACE. In the infantry service, generally recko which the soldier is trained to take steadily. 75 paces are taken in a minute; in quick time, 10 120 paces, the outward file stepping 33 inc quick time, 150 paces of 36 inches, or one y minute.

PALISADES. Triangular prismatic beams of w inches on each side, sharpened at the top and i three feet into the ground at intervals of s They are placed in the covered way 3 feet from the crest of the glacis, to secure it from surpri are generally about a foot above the crest of t hind which they are placed, and 8½ feet abo round each traverse when there is no banquette

PARADE. To assemble troops in a uniform manner, of regular muster, exercise, and inspection. I also the ground on which the exercises are

PARALLELS. Deep and wide trenches, generally th connecting the several lines of attack of a besi each other. The first is about 600 yards from t the second 300, and the third near or on the cres

PARAPET. Generally a mass of earth raised on the exterior crest of the rampart next the enemy, 18 or 20 feet broad, and 7 or 8 feet high, to cover the troops behind it from the fire of the besiegers. Also, generally, banks thrown up to cover and assist the defence of a position.

PARK OF ARTILLERY. The whole train of artillery belonging to an army.

PATROLE. A small party of men under the charge of a subaltern or non-commissioned officer, detached from the guard, to keep moving along streets or roads, to maintain the order and regularity of troops, &c. Patroles are also sent out to gain intelligence of the position and force of an enemy. This duty requires great caution and activity.

PERCUSSION CAPS. Small caps of copper filled partially with an explosive composition, which is fired when struck smartly with considerable force between two portions of hard metal, as steel. These now almost entirely supersede the use of flints in exploding fire-arms.

PETARD. A large vessel or machine of gun metal secured to a strong square-board, having iron hooks attached to it, to fix it against gates or palisades. This kind of pot is filled with 8 or 10 pounds of gunpowder, which, being fired, destroys the objects before it, and procures an entrance for an enemy. Leather or strong canvass bags are also sometimes employed in cases of emergency, which are more expeditious, and equally successful.

PIONEERS. Soldiers trained to work with various tools or instruments, such as pickaxes, hatchets, saws, spades, &c. Each company of a regiment furnishes one man to complete this body, formed under the command of a corporal. Their services are very important in clearing forests, working in intrenchments, completing approaches, and forming mines.

PIQUET. A detachment, formed either of infantry or cavalry, sent out for the purpose of guarding an army from surprise, &c.

PITCHING AND SCENDING. The movement of a ship, by which she plunges her head and stern alternately in the hollow of the sea.

PIVOT. The officer or soldier stationed at the flank on which a company wheels.

PLACES OF ARMS. Spaces at the salient and entrant angles of the covered way. The salient places of arms are the positions at which troops destined for a sortie generally assemble. The entrant places of arms contain troop various branches of the covered way. the places of arms are frequently furn the purpose of greater strength and se

PLATFORM. A floor, generally construc cannon are placed, behind an embrasu

POINT-BLANK. The position of a gun or the bore or barrel, and the objects aimed either horizontal or inclined. The poin is the distance that a shot is project stances, till it strikes the ground.

PONTOON. A species of boat for construc the passage of an *unfordable* river by approved were invented by General I shard of the Royal Engineers. T frequently reckoned the more conve copper with traverses, and by that me coarse weather. They are moored i distances, and connected with beams troops, artillery, and the stores of an readily.

PORT. A name given, on some occasion side of a ship, especially in steering, to ing lafboard for starboard, which in dif dangerous.

POSTERN. A passage constructed under communication from the fort into the

POUCH. A case of strong leather, gener divisions, for the purpose of carrying It is covered by a flap, to preserve the

RAKE. To cannonade a ship, at the h shot may scour the whole deck.

RAMPART. A broad embankment of ea stone or brick, surrounding a fortified main-works, commonly called the *once

RANGE. The distance from the piece fir shot strikes the ground.

RANK AND FILE. All those soldiers wl carry muskets.

RATION. An allowance of provisions ge

RAVELIN. A detached work composed salient angles, and raised before the c

**Rear-guard.** A detachment of troops appointed to protect the rear of an army.

**Reconnoitre.** To examine a country, so as to acquire a familiar knowledge of it, and to supply the defects of maps, required chiefly for military purposes. Particular parts of the map are distinguished by general marks of reference, connected with a minute memoir in writing.

**Redoubt.** A small work, frequently square, without bastions, placed at some distance from a fortification, to guard a pass, or obstruct the progress of an enemy in a given direction. A redoubt is sometimes made circular, because it will contain the greatest number of troops in a given space, and afford a superior defence. Redoubts have generally ditches and some means of giving a flanking fire, especially to those parts likely to be attacked.

**Reduce.** To compel a garrison to surrender by means of attack.

**Re-entering,** or simply *entering angle.* An angle in the fortification, pointing inwards to the place, and used in contradistinction to salient.

**Refuse.** In military operations, to throw or keep back the troops, to avoid an engagement with an enemy advantageously placed.

**Rendezvous.** A place appointed for the assembling of an army or of any body of troops.

**Reserve.** A select body of troops, kept in the rear for some particular object, such as to support an attack, or final charge, to terminate a battle successfully.

**Reveillé.** The beat of drums at the break of day. After this the sentries do not challenge.

**Revetement.** An exterior wall, or facing of stone or brick, supporting the front of the rampart on the side of the ditch.

**Ricochet.** A particular mode of firing guns at a low angle and loaded with a small charge. In this case the shot just clears the parapet of the enemy's works, and, rolling along the rampart, destroys the guns and kills numbers of men.

**Riflemen.** Light infantry, armed with rifles instead of muskets, trained to be expert marksmen, and having a peculiar drill and exercise of their own.

**Rocket.** A species of firework, frequently used for signals. Congreve rockets are most destructive missiles, frequently containing a shell, and very effectively employed in the attack of fortified places, in the destruction of shipping, and in various other warlike operations.

**Rolling.** A ship's motion from side to side, at r pitching.

**Running fire.** When troops fire rapidly in succe sition to a general discharge from the whole line.

**Sack.** To storm a town and pillage it.

**Salient angle,** in fortification. An angle proje country.

**Sally.** A secret and offensive movement of a st troops from a besieged place, in order to destroy the besiegers.

**Sally-ports.** Openings in the glacis, of 10 or 12 fe afford free egress and ingress to troops engaged sortie.

**Sand bags.** Bags of earth employed to repair bre brasures. The smaller kind are used on a parapet, make a small opening, like a loop-hole, by placing distance, and covering it with the third.

**Sentry or sentinel.** A soldier placed in a positio motions of an enemy, to prevent surprise, or t orders as he may be intrusted with.

**Sergeant.** A non-commissioned officer, selected porals, on account of his intelligence, steadiness good conduct, and intrusted with several responsi

**Sergeant-major.** The highest non-commissioned regiment, and, from the nature of his duties, i gree an assistant to the adjutant.

**Colour-sergeant.** A non-commissioned officer wl attend the colours in the field. This office is distinction, given only to men of valour and fidel

**Shaft.** In mining, a perpendicular excavation, exte depth, from which the several branches of a min

**Shrapnel-shells,** or Spherical Case-shot, have a than common case or canister, and are very eff fare.

**Siege.** The art of surrounding a town or fortifi besieging army, and attacking it with artiller mines and trenches, so as to destroy the princ and ultimately to storm the place, unless it yiel capitulation.

**Skirmish.** A kind of irregular engagement between in presence of both armies, for the purpose of c movements of troops, or bringing on a general bi

SLIP A CABLE. To allow the whole cable attached to an anchor to run out, when, on account of all possible expedition being required, there is not time to weigh it in the usual manner.

SPIKE. To *spike* guns is to drive large nails into the vent, or touch-hole, to render them unserviceable for some time, till it can be cleared.

SQUADRON. In the army is a body of cavalry consisting of two troops. A *Squadron* in the navy is a small fleet not exceeding five ships.

SQUARE. A form into which infantry are thrown to resist a charge of cavalry—the soldiers all facing outwards, whatever their number.

STARBOARD. The right side of the ship to a spectator looking forwards.

STAR FORTS. Forts with several salient angles, in the form of a star, as generally represented.

STORES, military and marine. Arms, ammunition, provisions, clothes, and other necessaries.

STRATEGY. The science of war.

STORM. To make a powerful and vigorous assault on any position occupied by an enemy.

SUBALTERN. All military officers under the rank of a captain.

SUTLER. A camp-follower, who sells drink and provisions to the troops.

TACTICS. The science and art of disposing military and naval armaments for battle, often comprehending the whole science of war, and the means prepared for carrying it on.

TENAILLE. A work usually constructed on the lines of defence in front of the curtain. In modern fortification it has two faces, in line with those of the adjacent bastions, with an intermediate curtain parallel to that of the main works.

TENAILLON. Formerly a small work placed on each side of the ravelin, for additional strength.

TOISE. An old French measure of 6 French feet. The English fathom bears to the French toise the ratio of 1 to 1·066 nearly. Hence, for rough purposes, the English fathom may be accounted nearly equal to the French toise.

TRAVERSES. Generally parapets of earth, formed in the covered way, to cover troops placed there from siegers. Their superior slopes are dir and are furnished with banquettes b next the crest of the glacis, there transit of troops, designated *en crochet* o to their shape. In the course of a whenever they may be required to co

TRENCHES. Ditches made during a sie to approach the works as securely a zig-zag shape, each portion cleari through each position.

TROOP. In the singular, generally sig goons, under the command of a captai

TROUS-DE-LOUP, or wolf-holes. Are dug cone, having a small picket, sharpene bottom.

TRUCKS of a gun-carriage. The wheels

TRUNNIONS of a gun. The cylindrical ar to its carriage. To render guns useles times struck off.

TUMBRILS. Covered carts, employed t tools of pioneers, miners, &c.

VAN of an army. The front or first lin

VAN, in the navy. The first division occasionally the leading ship.

VIDETTES or VEDETTES. Sentries placed points, so as to be able to observe adv of an enemy, and to give early notice

UNDER ARMS. The condition of troo armed and accoutred.

WEAR or VEER. To change a ship's cours by turning her stern to windward. T practised, when that of tacking, by wind, would be dangerous.

WEATHER-GAUGE. When one ship is t is said to have the weather-gauge of l

*To* WINDWARD is towards that part of the wind blows.

WINGS of an army. The extreme right

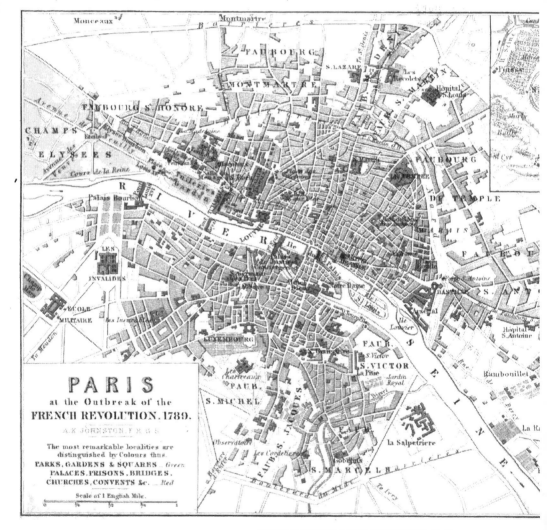

PARIS
at the Outbreak of the
FRENCH REVOLUTION, 1789.

A. K. JOHNSTON, F.R.G.S.

The most remarkable localities are
distinguished by Colours thus:
PARKS, GARDENS & SQUARES ... *Green*
PALACES, PRISONS, BRIDGES,
CHURCHES, CONVENTS &c. ... *Red*

Scale of 1 English Mile.

WILLIAM BLACKWOOD & SONS, EDINBURGH & LONDON.

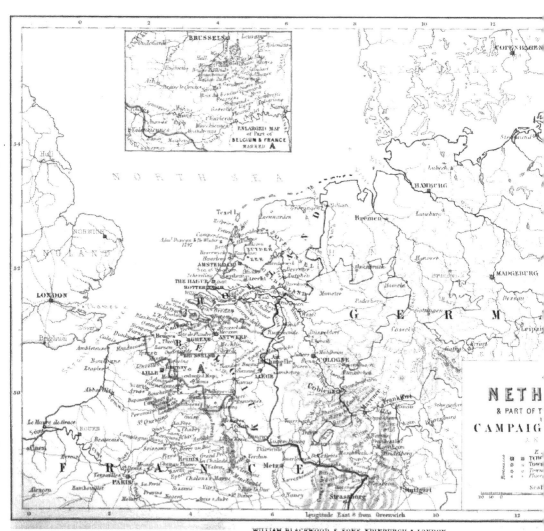

ENLARGED MAP
of Part of
BELGIUM & FRANCE
MARKED A

NORTH SEA

COPENHAGEN

LONDON

ENGLAND

AMSTERDAM
THE HAGUE
ROTTERDAM

ANTWERP
BRUSSELS

HAMBURG

Bremen

GERMANY

MADGEBURG

Leipzig

COLOGNE

LIEGE

Coblentz

NETH
& PART OF T
CAMPAIG

FRANCE

PARIS

Reims
Metz

Strasbourg

Stuttgart

WILLIAM BLACKWOOD & SONS, EDINBURGH & LONDON.

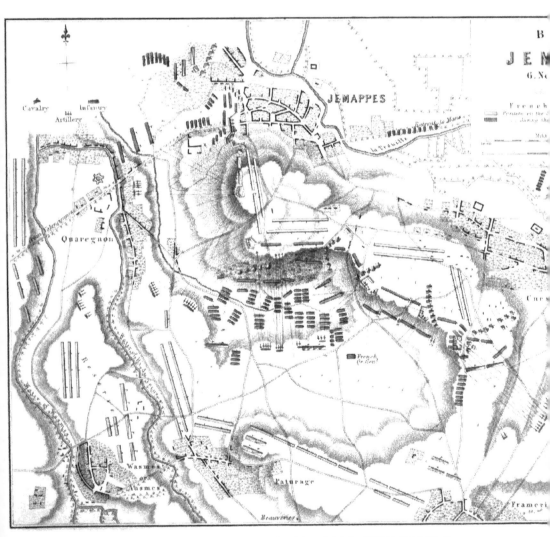

BATTLE OF
JEMAPPES
6. November 1792

French
Prussian on the...
...during the...

JEMAPPES

Cavalry    Infantry
Artillery

Quaregnon

French
Gen¹

Cuesmes

Wasmes
or
Jusmes

Paturage

Frameri

Beauvirage

WILLIAM BLACKWOOD & SONS EDINBURGH & LONDON

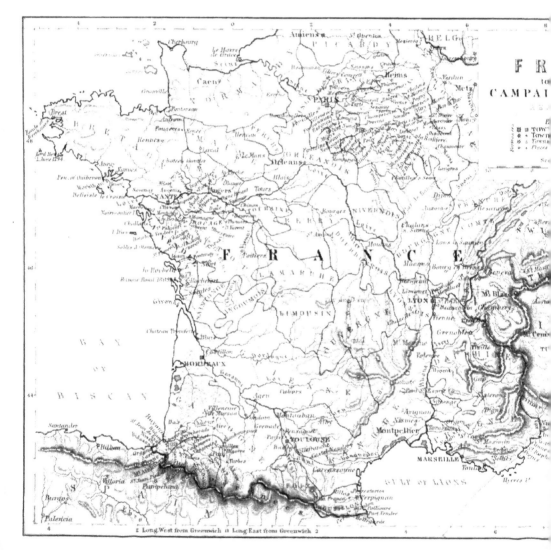

WILLIAM BLACKWOOD & SONS, EDINBURGH & LONDON

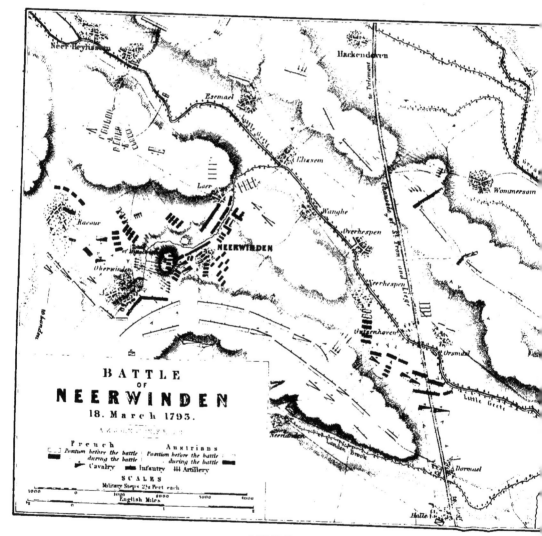

BATTLE
of
NEERWINDEN
18. March 1793.

French     Austrians
Position before the battle   Position before the battle
during the battle    during the battle
Cavalry    Infantry   Artillery

SCALES
Military Steps 2½ Feet each
1000   0   1000   2000   3000   4000
English Miles

WILLIAM BLACKWOOD & SONS, EDINBURGH & LONDON.

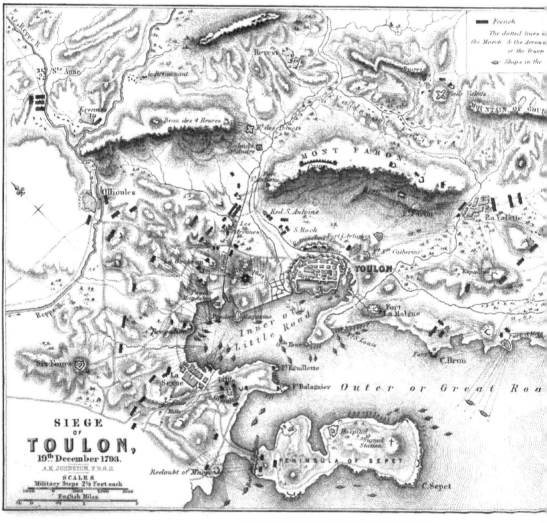

SIEGE
OF
TOULON,
19th December 1793.
A. K. JOHNSTON, F.R.G.S.
SCALES
Military Steps 2½ Feet each
English Miles

WILLIAM BLACKWOOD & SONS, EDINBURGH & LONDON.

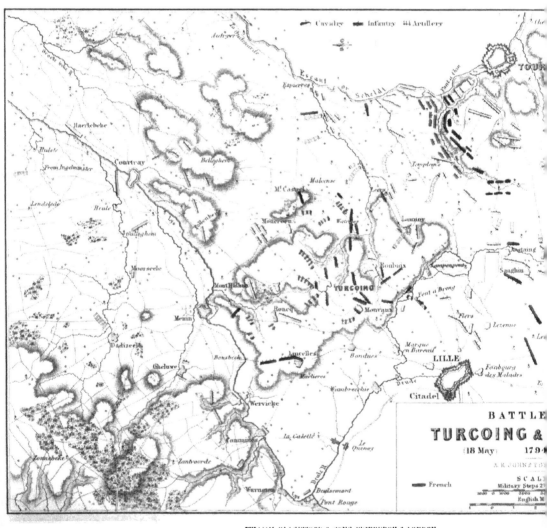

BATTLE
TURCOING &
(18 May)    1794

A.K.JOHNSTON

SCALE
Military Steps 2?

English M

WILLIAM BLACKWOOD & SONS, EDINBURGH & LONDON

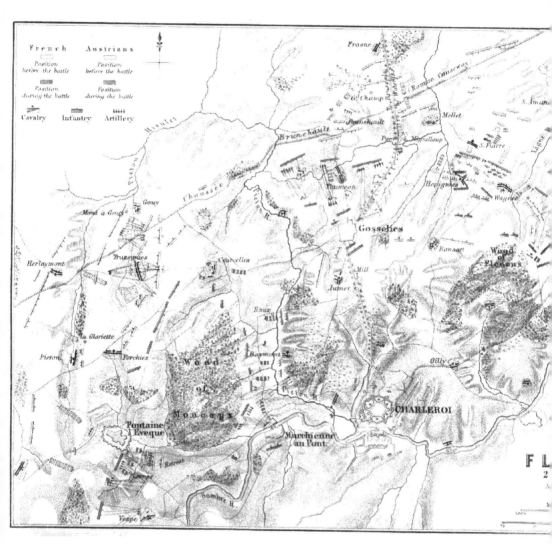

WILLIAM BLACKWOOD & SONS EDINBURGH & LONDON

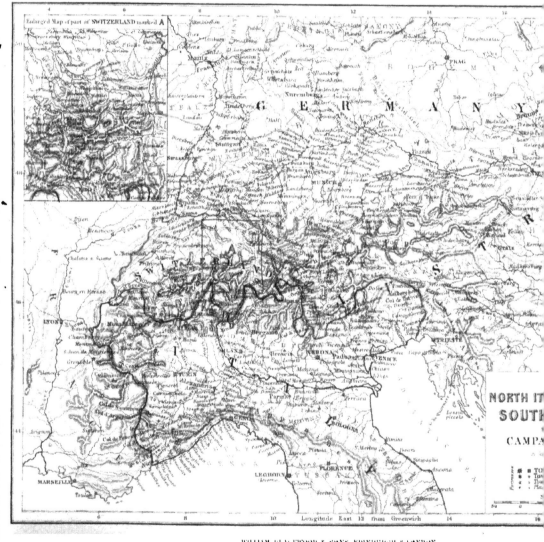

Enlarged Map of part of SWITZERLAND marked A

NORTH IT...
SOUTH...
CAMPA...

WILLIAM BLACKWOOD & SONS EDINBURGH & LONDON

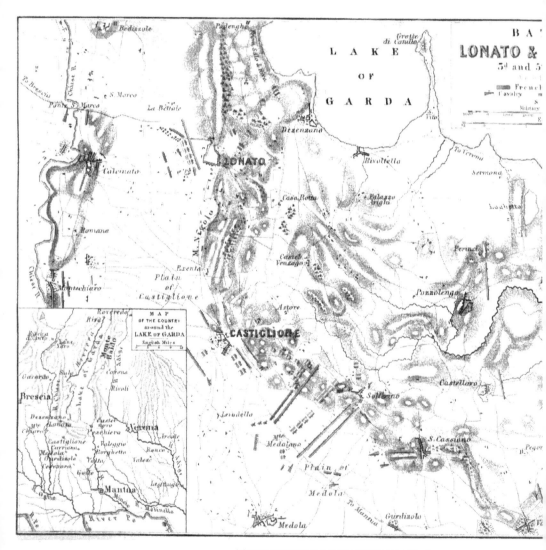

WILLIAM BLACKWOOD & SONS EDINBURGH & LONDON

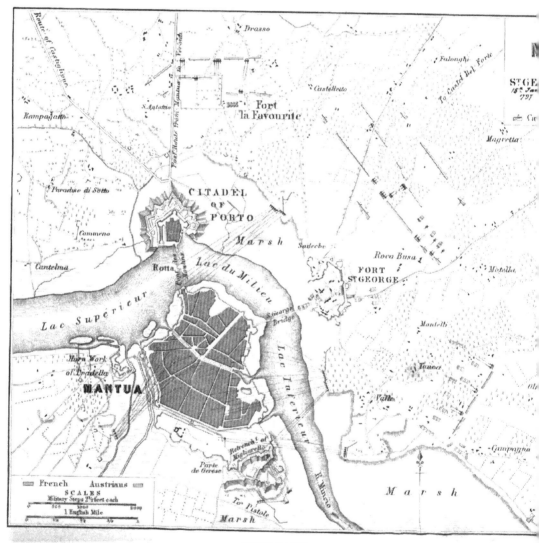

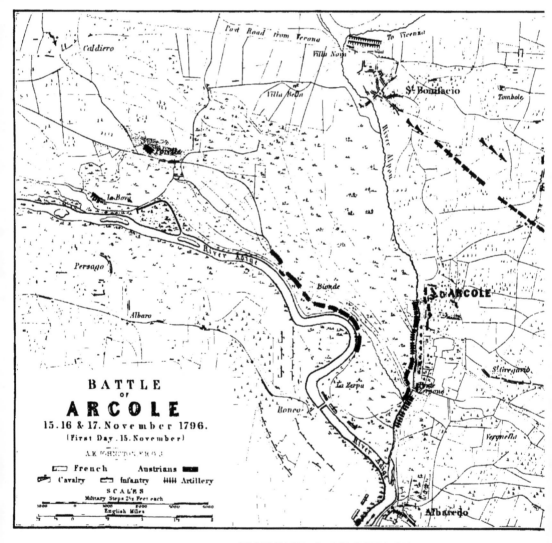

Caldiero

Post Road from Verona

Villa Nova

To Vicenza

St Bonifacio

Tombole

Villa Bella

Porcile

Der la Bova

River Alpon

River Alpo

Persago

Bionde

D'ARCOLE

Albaro

St Gregorio

La Zerpa

Porto Vergano

Ronco

River Adige

Vergnella

# BATTLE
## OF
## ARCOLE
### 15. 16 & 17. November 1796.
#### (First Day . 15. November)

A.K JOHNSTON F.R.G.S.

French        Austrians

Cavalry        Infantry        Artillery

## SCALES
Military Steps 2½ Feet each

1000        2000        3000        4000

English Miles

Albaredo

WILLIAM BLACKWOOD & SONS. EDINBURGH & LONDON

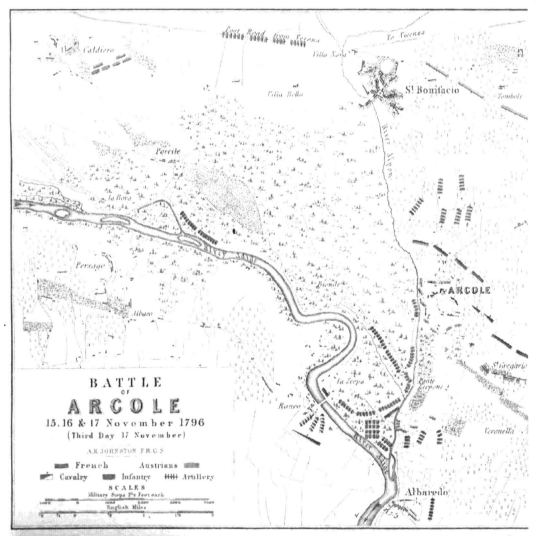

BATTLE
OF
ARCOLE
15. 16 & 17 November 1796
(Third Day 17 November)

A.K. JOHNSTON F.R.G.S

French        Austrians
Cavalry    Infantry    Artillery

SCALES
Military Steps 2's Feet each

English Miles

WILLIAM BLACKWOOD & SONS EDINBURGH & LONDON

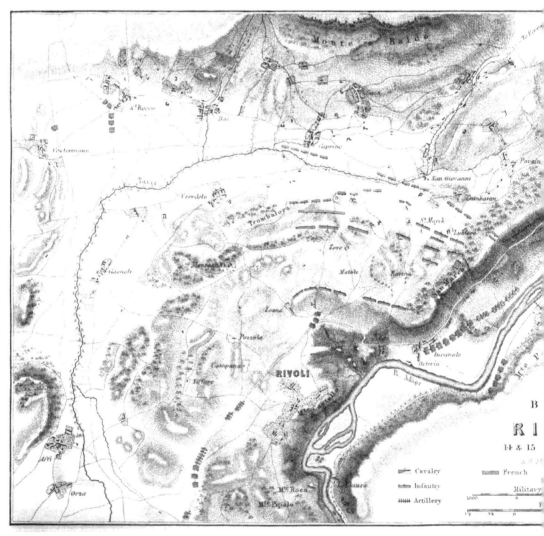

WILLIAM BLACKWOOD & SONS EDINBURGH & LONDON

Patten  CAMPERDOWN

D U T C H   F L E E T

Galatea  Daphne  Ajax  Waakzaamheid  Minerva  Brister  Innocent  Batavia

Beschermer  Gelykheid  Hercules  Devries  VRYHEID  States General  Wassenaer  Batavier  Brutus  Leyden  Mars  Cerberus  Jupiter

Monarch  Powerful  Montague

Wind S.S.W.

VENERABLE

Triumph

Ardent

Bedford

Acton  Martin

Lancaster  Director  Circe  King George  Rose

Diligent

Belliqueux

Speculation

Adamant  Isis

## BATTLE
## of
## CAMPERDOWN
### 11th October 1797.

■ British  Dutch ■

B R I T I S H   F L E

WILLIAM BLACKWOOD & SONS, EDINBURGH & LONDON.

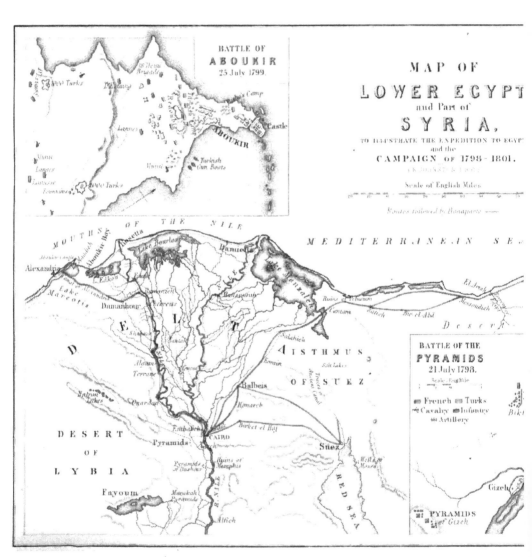

BATTLE OF
**ABOUKIR**
25 July 1799.

MAP OF
LOWER EGYPT
and Part of
SYRIA,
TO ILLUSTRATE THE EXPEDITION TO EGYPT
and the
CAMPAIGN OF 1798-1801.

Scale of English Miles

Routes followed by Bonaparte

MEDITERRANEAN SEA.

MOUTHS OF THE NILE

Alexandria

Lake
Mareotis

D E L T A

I S T H M U S
OF SUEZ

Damanhour

Balbeis

D E S E R T
OF
L Y B I A

Pyramids

CAIRO

Suez

R E D   S E A

Fayoum

NILE

Desert

BATTLE OF THE
PYRAMIDS
21 July 1798.

French Turks
Cavalry Infantry
Artillery

PYRAMIDS
of Gizeh

Gizeh

WILLIAM BLACKWOOD & SONS, EDINBURGH & LONDON.

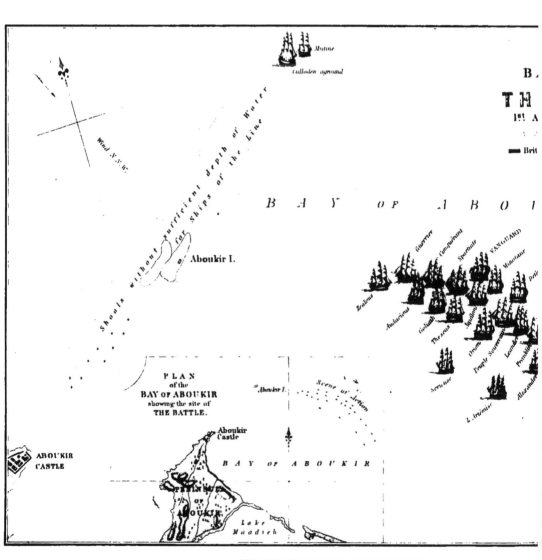

Mutine

Culloden aground

B A

T H

1st A

Brit

*B A Y   O F   A B O U*

Aboukir I.

Guerrier   Conquerant   Spartiate   VANGUARD

Minotaure

Zealous

Audacious

Goliath   Squadron

Orion

Theseus   Serieuse

Peuple Souverain   Leander

Franklin

Alexandre

Serieuse

L. Artemise

*Wind N.N.W.*

*Shoals without sufficient depth of Water for Ships of the Line*

P L A N
of the
**BAY OF ABOUKIR**
showing the site of
**THE BATTLE.**

Aboukir I.

Scene of Action

Aboukir
Castle

*B A Y   of   A B O U K I R*

**ABOUKIR
CASTLE**

POINT
OF
ABOUKIR

Lake
Maadieh

WILLIAM BLACKWOOD & SONS, EDINBURGH & LONDON

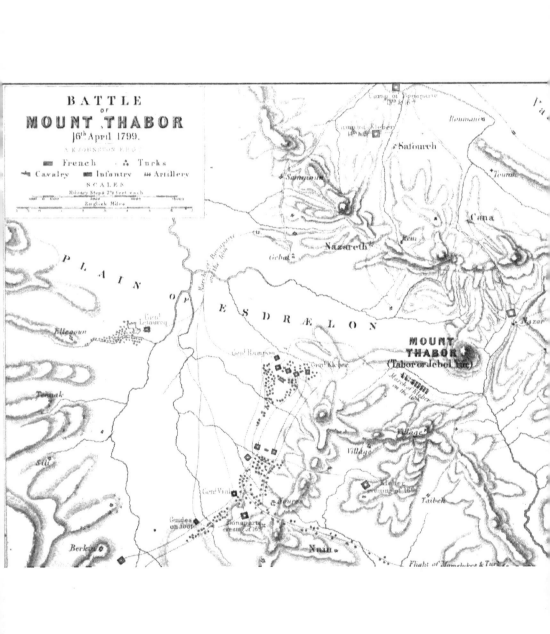

BATTLE
OF
MOUNT THABOR
16th April 1799.
A. K. JOHNSTON F.R.G.S.

French ▬  Turks ▲
Cavalry ✦  Infantry ■  Artillery ⚏

SCALES
Military Steps 2½ feet each
English Miles

Camp of Bonaparte
Renmane
Camp of Kleber
Safoureh
Summanin
Cana
Rem
Nazareth
Gebat
Ellegaun
Genl Letourcq
PLAIN
OF
ESDRAELON
Nazar
Genl Rampon
MOUNT
THABOR
(Tabor or Jebel Tur)
Genl Kleber
March of Kleber on the 15th
Temak
Village
Sili
Village
Genl Viale
Kleber encamping on the 15th
Taibeh
Guides on foot
Bonaparte
evening of 16th
Berkoun
Nain
Flight of Mamelukes & Turks

BATTLE
OF
STOCKACH
25. March 1799.

A.K.JOHNSTON, F.R.G.S

French ▬▬▬ Austrians ▬▬▬
Cavalry ▬▬  Infantry ▬▬  Artillery ▬▬

SCALES
Military Paces 2½ Feet each

English Miles

WILLIAM BLACKWOOD & SONS, EDINBURGH & LONDON.

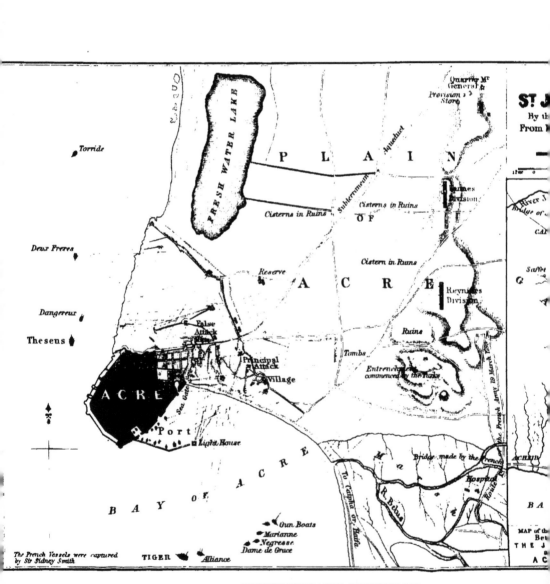

Torride

Deux Preres

Dangereux

Theseus

FRESH WATER LAKE

P L A I N

Quartier Mr General

Provision Store

Aqueduct

Cisterns in Ruins

Subterranean

Cisterns in Ruins

O F

Kleines Division

Cistern in Ruins

Reserve

A C R E

Reynier's Division

False Attack

Principal Attack

Village

Ruins

Tombs

Entrenchment commenced by the Turks

A C R E

Port

Sea Gate

Light House

M a

Bridge made by the French

To Caipha or Haiffa

R. Belus

Hospital

French Army 19 March 1799

ACHRIB

B A Y   O F   A C R E

B A

Gun Boats

Marianne

Negresse

Dame de Grace

*The French Vessels were captured by Sir Sidney Smith.*   TIGER   Alliance

ST J

By the

From

River J

Bridge of

CAI

Saffet

Q

MAP of the

Bet

THE J

a

A C

WILLIAM BLACKWOOD & SONS, EDINBURGH & LONDON.

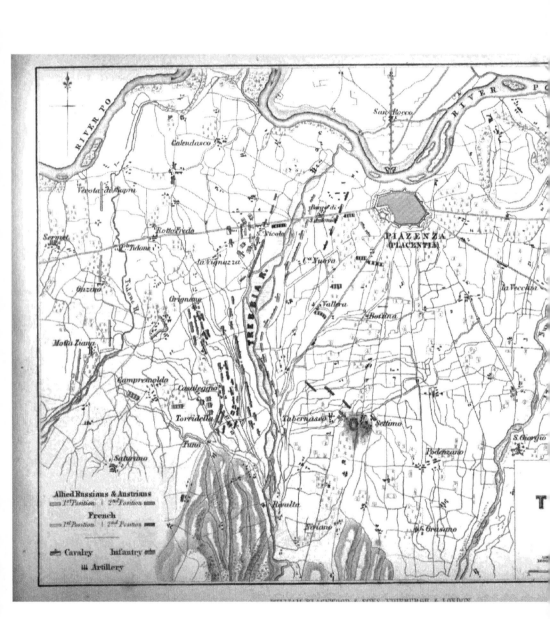

RIVER PO

RIVER PO

San Rocco

Calendasco

Verola di Supra

PIAZENZA
(PLACENTIA)

Sergué
Rotto Freda

S.<sup>ta</sup> Tidone

S. Nicolo

la Vignuzza

C.<sup>a</sup> Nuova

la Vecchia

Guzano

TREBIA R.

Grignano

Vallera

Botura

Tidone R.

Motta Ziana

Campremoldo

Casoliraguo

Torridella

Tabernasso

Settimo

S. Giorgio

Tuna

Podenzano

Satarano

Allied Russians & Austrians
═══ 1<sup>st</sup> Position ┊ 2<sup>nd</sup> Position ═══
French
═══ 1<sup>st</sup> Position ┊ 2<sup>nd</sup> Position ═══

Rivalta

Cavalry      Infantry
Artillery

Siviano

Grusano

T

WILLIAM BLACKWOOD & SONS, EDINBURGH & LONDON

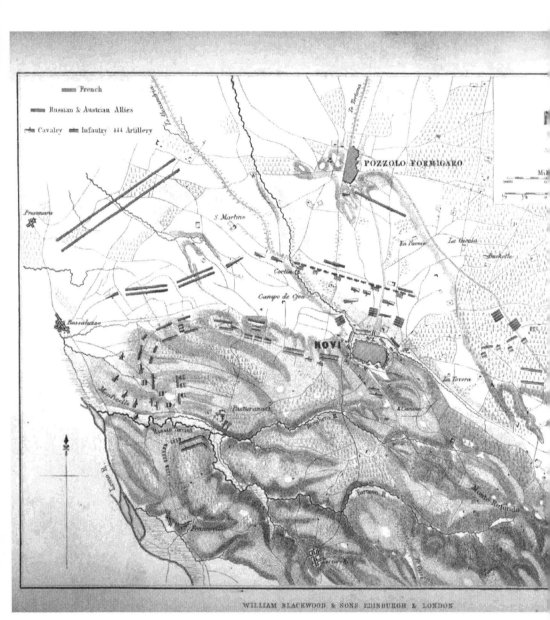

French

Russian & Austrian Allies

Cavalry   Infantry   Artillery

Presnara

S. Martino

POZZOLO FORMIGARO

To Tortona

La Ravina

La Girola

Buchetta

Cortia G.

Campo de Cjoa

Bassaluzao

NOVI

La Fovera

Monte

Pasturanea

A Cavena

Monte Rotondo

Franconilla

BATTLE
of
ZÜRICH
4 June 1799.
AK JOHNSTON F.R.G.S.

French ——— Austrians ▦▦▦
🐎 Cavalry ▬ Infantry ▥▥ Artillery
SCALES
Military Steps 2¼ feet each
English Miles

WILLIAM BLACKWOOD & SONS. EDINBURGH & LONDON.

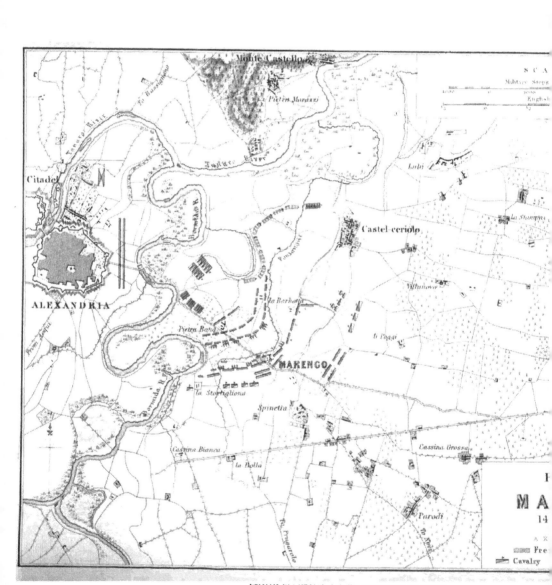

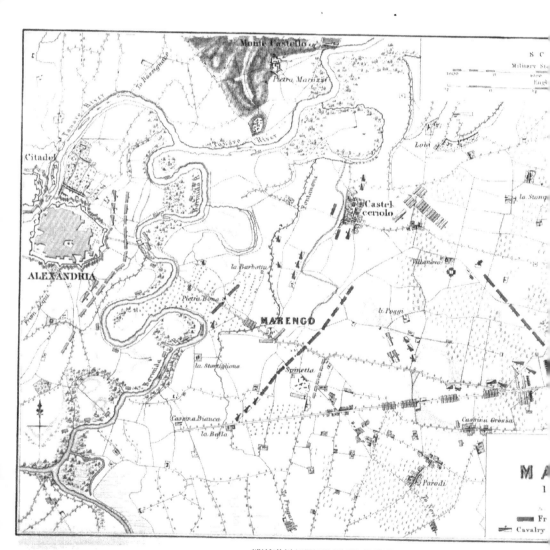

Monte Castello

Pietra Marazzi

Tanaro River

la Bossignana

Citadel

Fosaro River

Lobi

la Stanzi

Castel ceriolo

Bormida R.

ALEXANDRIA

la Barbotta

Villanava

Pietra Bona

MARENGO

la Poggi

Fram Lagia

la Stanziliano

Spinetta

Cassina Bianca

la Bella

Cassina Grossa

Bermida

M A

Parodi

Fr

Cavalry

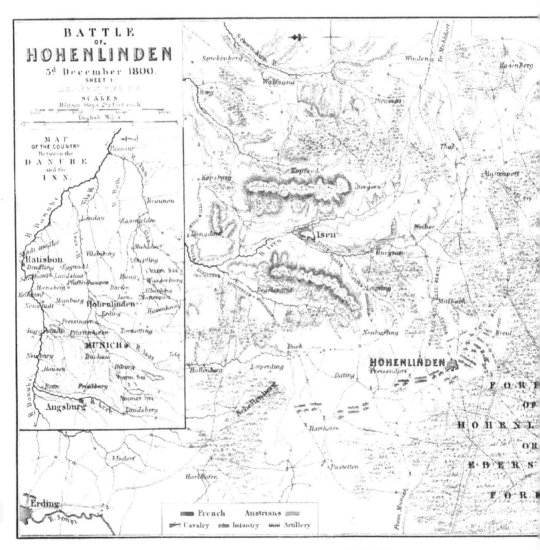

BATTLE
OF.
HOHENLINDEN
5ᵈ December 1800.
SHEET I

SCALES
Military Steps 2½ feet each
English Miles

MAP
OF THE COUNTRY
Between the
DANUBE
and the
INN.

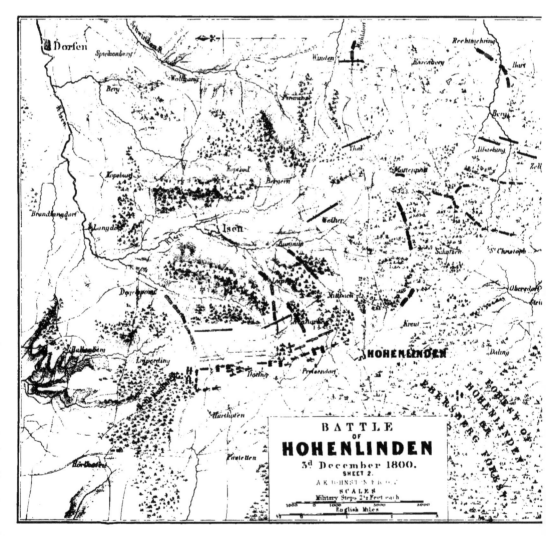

BATTLE
OF
HOHENLINDEN
3ᵈ December 1800.
SHEET 2.

A K JOHNSTON F R G S

SCALES
Military Steps 2½ Feet each
English Miles

WILLIAM BLACKWOOD & SONS EDINBURGH & LONDON

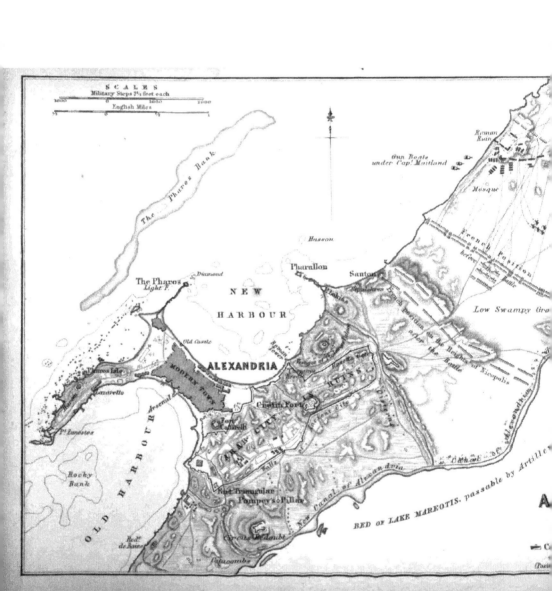

WILLIAM BLACKWOOD & SONS, EDINBURGH & LONDON.

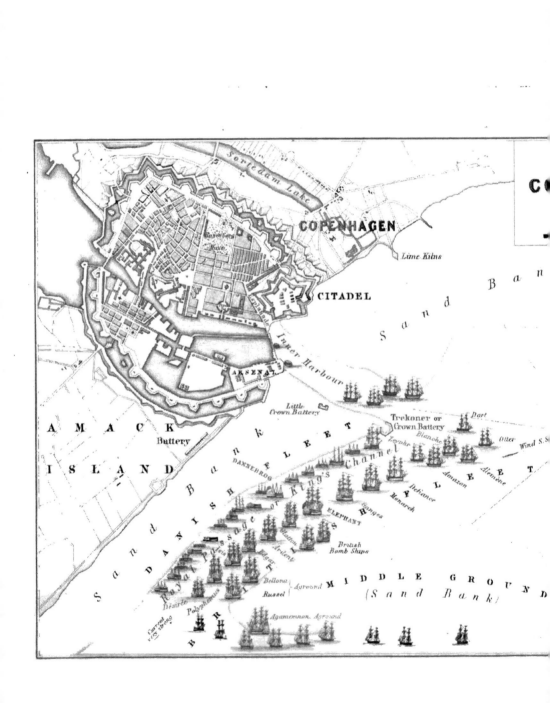

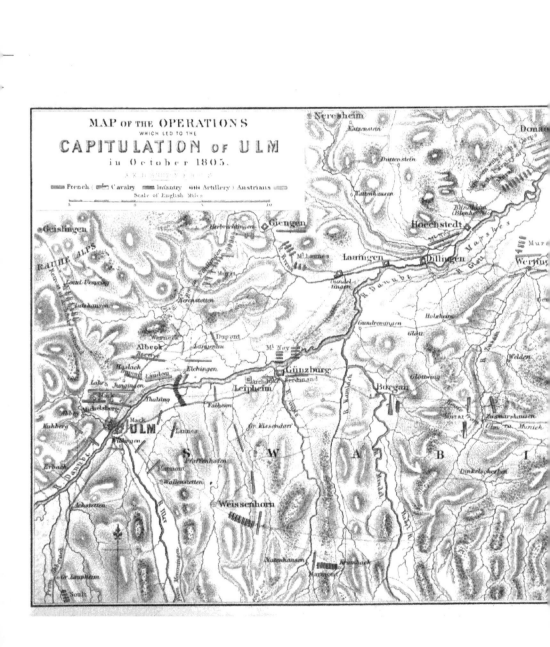

MAP OF THE OPERATIONS
WHICH LED TO THE
CAPITULATION OF ULM
in October 1805.

French : Cavalry : Infantry : Artillery : Austrians
Scale of English Miles

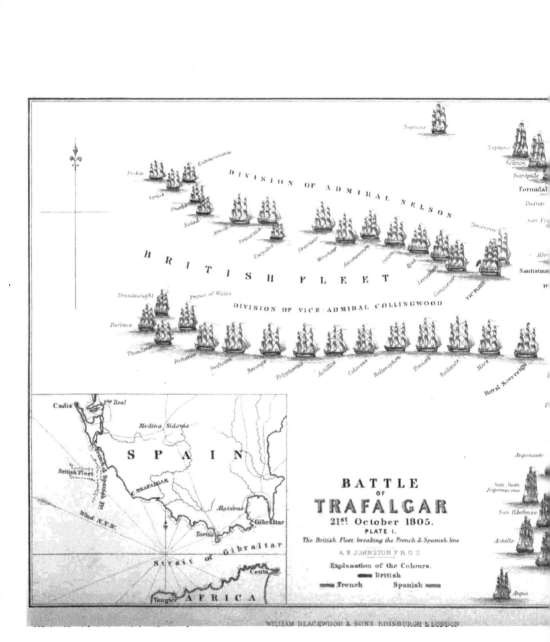

BATTLE
OF
TRAFALGAR
21ˢᵗ October 1805.
PLATE I.
The British Fleet breaking the French & Spanish line

A K JOHNSTON F R G S

Explanation of the Colours.
British
French    Spanish

WILLIAM BLACKWOOD & SONS EDINBURGH & LONDON

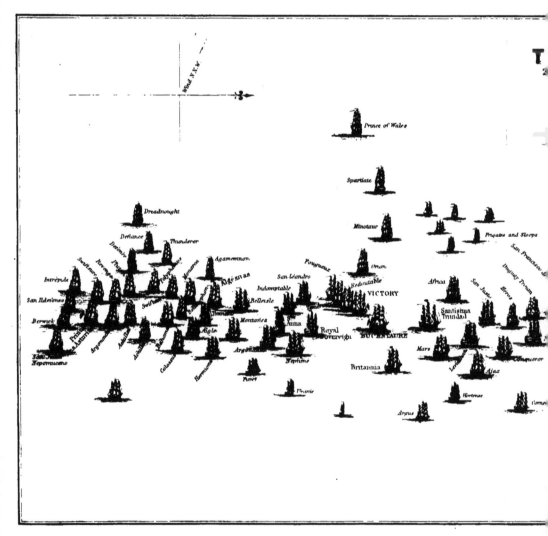

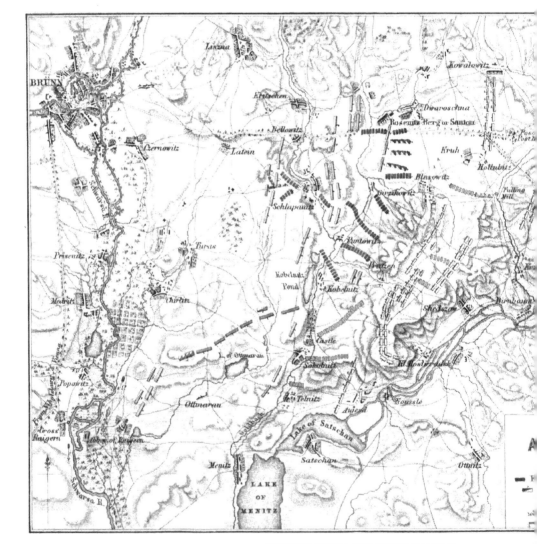

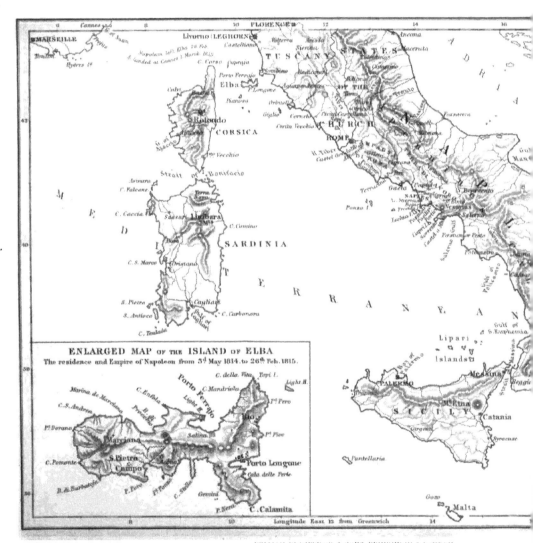

ENLARGED MAP of the ISLAND of ELBA
The residence and Empire of Napoleon from 3ᵈ May 1814 to 26ᵗʰ Feb. 1815.

WILLIAM BLACKWOOD & SONS, EDINBURGH & LONDON.

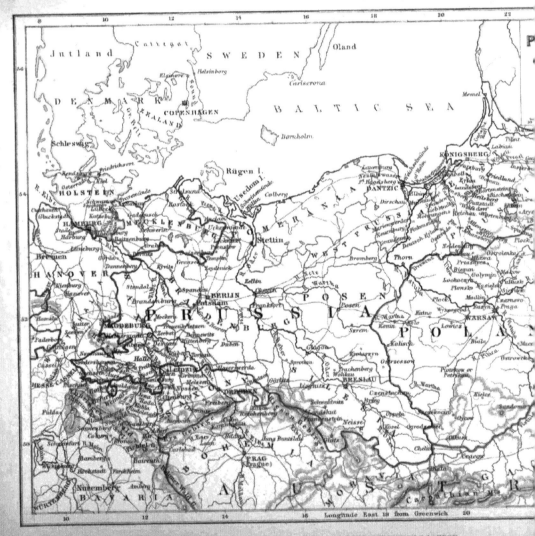

WILLIAM BLACKWOOD & SONS, EDINBURGH & LONDON.

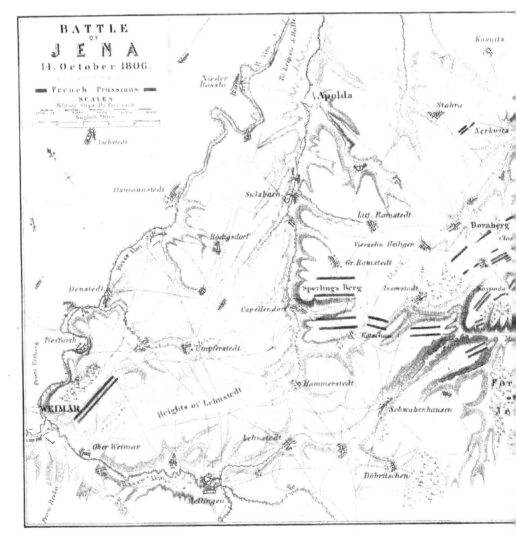

BATTLE
OF
JENA
14. October 1806.

French    Prussians
SCALES
Military Strips 2½ Feet each
English Miles

Nieder Rossle

Lehstedt

Osmannstedt

Rödigsdorf

Denstedt

River Ilm

Rethforth

Umpferstedt

WEIMAR

Ober Weimar

Heights of Lehnstedt

Mellingen

Sulzbach

Capellendorf

Lehnstedt

Apolda

Stohra

Nerkwitz

Kosnitz

Litt. Romstedt

Vierzehn Heiligen

Dornberg

Gr. Romstedt

Sperlings Berg        Assowstedt        Kospoda

Kotschau

Hammerstedt

Schwabenhausen

Döbritschen

WILLIAM BLACKWOOD & SONS EDINBURGH & LONDON.

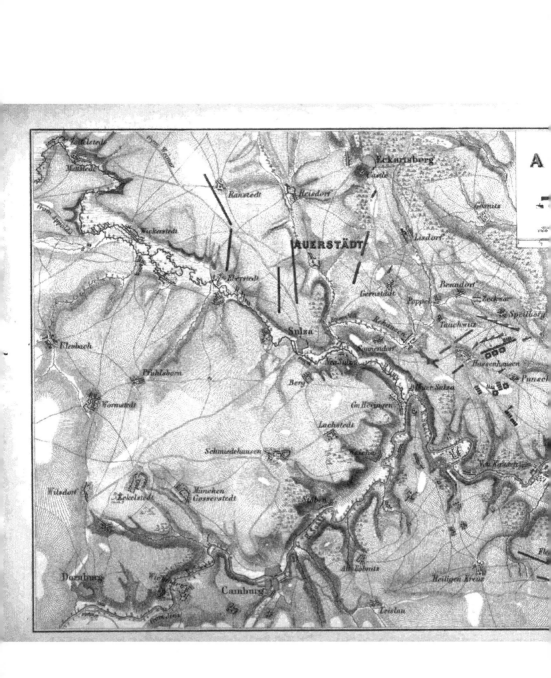

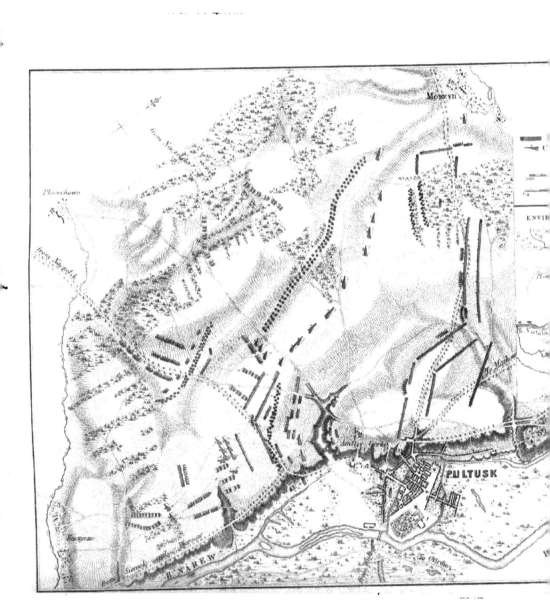

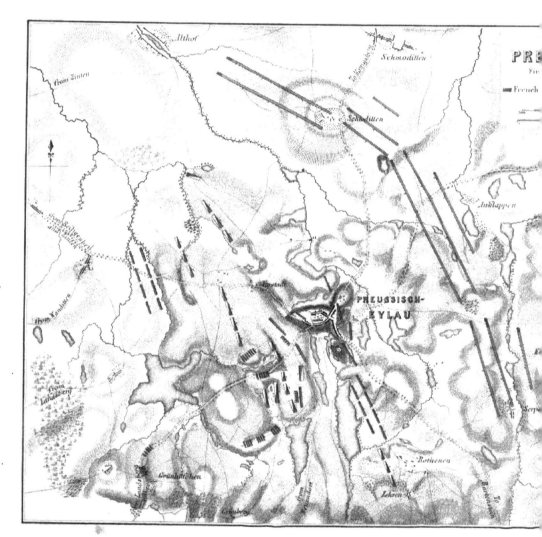

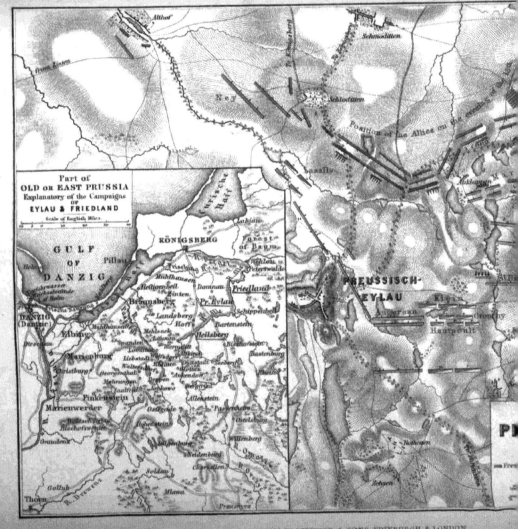

from Linden

Althof

Schmolitten

Pr. Eylau

Schlodtten

Position of the Allies on the

Lasalle

Auklappen

Part of
**OLD or EAST PRUSSIA**
Explanatory of the Campaigns
OF
**EYLAU & FRIEDLAND**
Scale of English Miles.

Labiau

Forest
of Baum

KÖNIGSBERG

Frisches Haff

**GULF**

Pischau

Rudau
Peterswalde

Pillau

Mühlhausen

Friedland

**DANZIG**

Hela

Heiligenbeil

Domnau

Zinten

Pr. Eylau

**PREUSSISCH-**

Schippenbell

**EYLAU**

Frischeshaff
of Balto

Braunsberg

**Klein**

DANZIG
(Danzic)

Landsberg

Hoff

Bartenstein

Anklappen

Grouchy

Dirschau

Mühlhausen

Mehlsack

Heilsberg

Hauptpoll

Liebenau
Wormditt

Elbing

Spanden
Londau

Bischofstein

**Marienburg**

Liebstadt
Waldau

Guttstadt
Deppen

Seeburg

Wolau

Ankendorf

Rhein

Christburg

Georgenthal

Bergfried

Mehrungen

Saalfeld
Jankowo

**Pinkenstein**

Allenstein

Bathenen

**Marienwerder**

Osterode

Passenheim

Guttstadt

Wilatschken
Blachofswerder

Hohenstein

Willenberg

Graudenz

Wolfenburg

Neidenburg

Tolzen

Golub

R. Drewenz

Soldau

Mlawa

Thorn

Prasznya

WILLIAM BLACKWOOD & SONS, EDINBURGH & LONDON.

BATTLE
OF
HEILSBERG
10ᵗʰ June 1807.

A.K.JOHNSTON F.R.G.S.

French ▬▬▬ Russians ▬▬▬
Cavalry ▬ Infantry ▬ Artillery ▥

SCALES
Military Steps 2½ feet each
English Miles

WILLIAM BLACKWOOD & SONS EDINBURGH & LONDON.

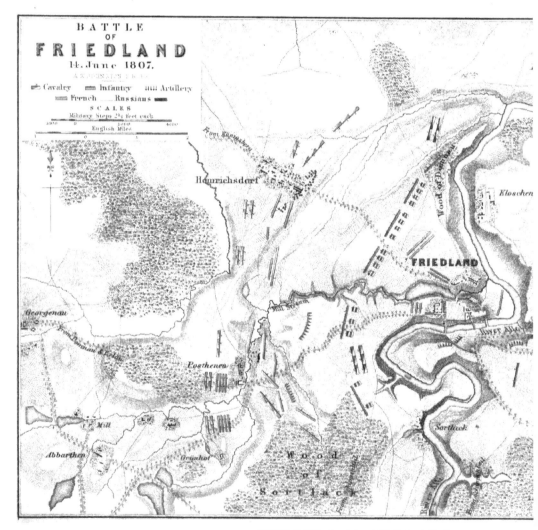

BATTLE
OF
FRIEDLAND
14. June 1807.

Cavalry    Infantry    Artillery
French    Russians

SCALES
Military Steps 2½ feet each
English Miles

Heinrichsdorf

Wood of D...

Kloschen

FRIEDLAND

Georgenau

Posthenen

Mill

Abbarthen

Grünhof

Wood
of
Sortlack

Sortlack

WILLIAM BLACKWOOD & SONS, EDINBURGH & LONDON

SIEGE OF
**SARAGOSSA,**
by the
FRENCH ARMY OF ARAGON;
in 1808 and 1809.
A & JOHNSTON FR.G.S.

*The Buildings destroyed by the Bombard
ment are distinguished by light colouring*

SCALE OF ¼ A MILE

Military Steps 2½ feet each

WILLIAM BLACKWOOD & SONS, EDINBURGH & LONDON.

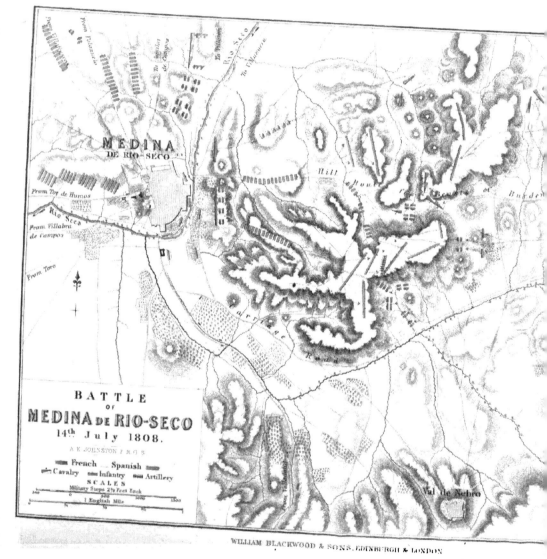

BATTLE
of
MEDINA DE RIO-SECO
14th July 1808.

A.K. JOHNSTON F.R.G.S.

French ▬▬  Spanish ▬▬
Cavalry   Infantry   Artillery
SCALES
Military Steps 2½ Feet Each
1 English Mile

WILLIAM BLACKWOOD & SONS, EDINBURGH & LONDON.

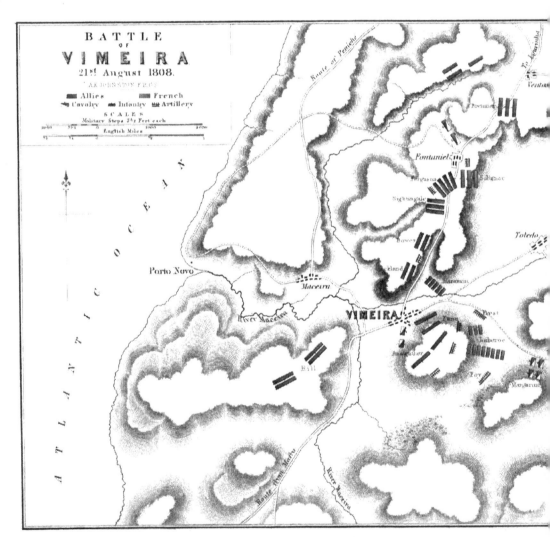

BATTLE
OF
VIMEIRA
21st August 1808.

BY AR JOHNSTON F.R.G.S

Allies    French
Cavalry  Infantry  Artillery

SCALES
Military Steps 2½ Feet each
English Miles

WILLIAM BLACKWOOD & SONS, EDINBURGH & LONDON.

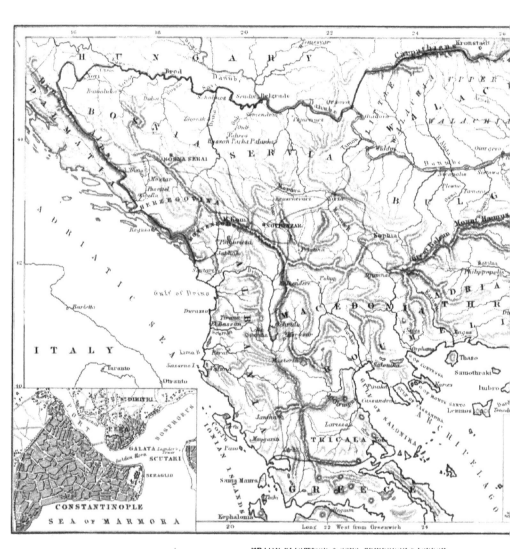

WILLIAM BLACKWOOD & SONS, EDINBURGH & LONDON.

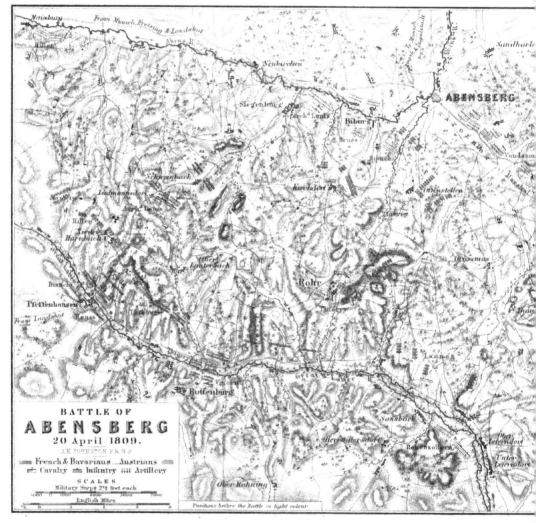

BATTLE OF
**ABENSBERG**
20 April 1809.
A K JOHNSTON F.R.G.S

French & Bavarians  Austrians
Cavalry  Infantry  Artillery

SCALES
Military Steps 2½ feet each

English Miles

*Positions before the Battle in light colour*

WILLIAM BLACKWOOD & SONS EDINBURGH & LONDON

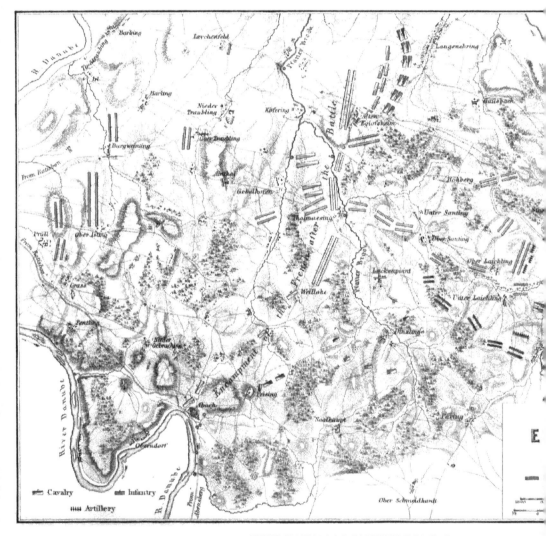

Cavalry  Infantry

Artillery

WILLIAM BLACKWOOD & SONS, EDINBURGH & LONDON.

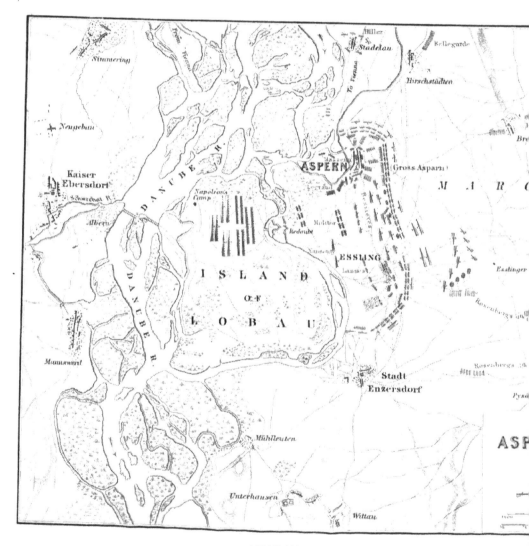

Simmering

Neugebau

Kaiser
Ebersdorf

Schwechat W.

Albern

Mannswört

DANUBE R.

DANUBE R.

Napoleon's
Camp

ISLAND

OF

LOBAU

Mühlleuten

Unterhausen

Redoubt

Molitor

Nausein

Lannes

ASPERN

Gross Aspern

ESSLING

Stadt
Enzersdorf

Wittau

Müller

Stadelau

Bellegarde

Hirschstädten

MARC

Esslinger

Rosenbergs

Rosenbergs

Pys

ASP

WILLIAM BLACKWOOD & SONS EDINBURGH & LONDON

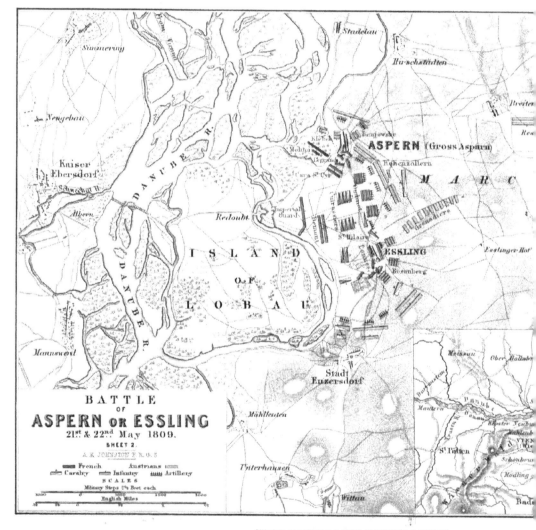

BATTLE
OF
ASPERN OR ESSLING
21st & 22nd May 1809.

SHEET 2.

A. K. JOHNSTON F.R.G.S.

French    Austrians
Cavalry    Infantry    Artillery

SCALES

Military Steps 2½ Feet each.

English Miles

WILLIAM BLACKWOOD & SONS, EDINBURGH & LONDON.

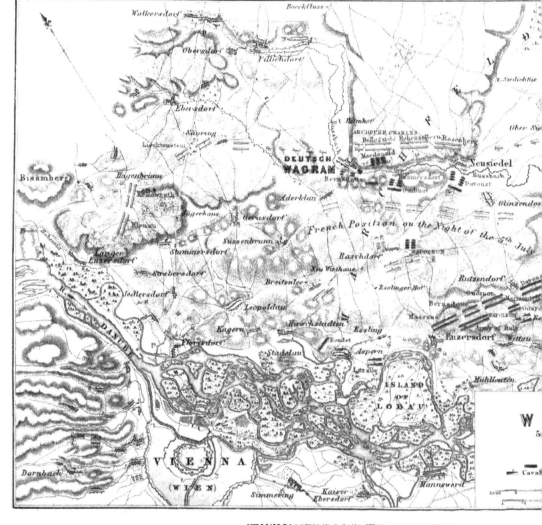

WILLIAM BLACKWOOD & SONS, EDINBURGH & LONDON.

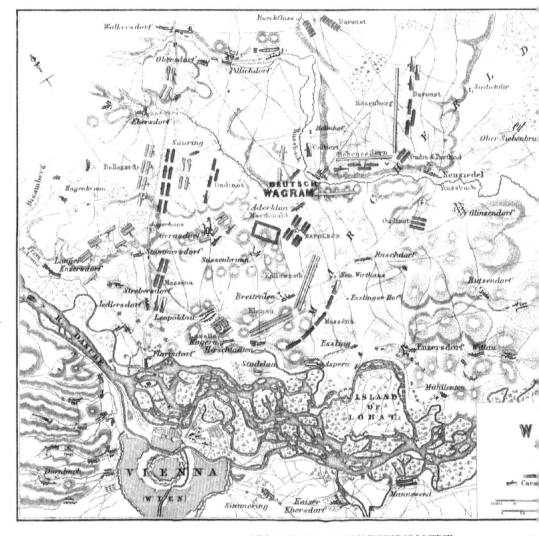

WILLIAM BLACKWOOD & SONS, EDINBURGH & LONDON.

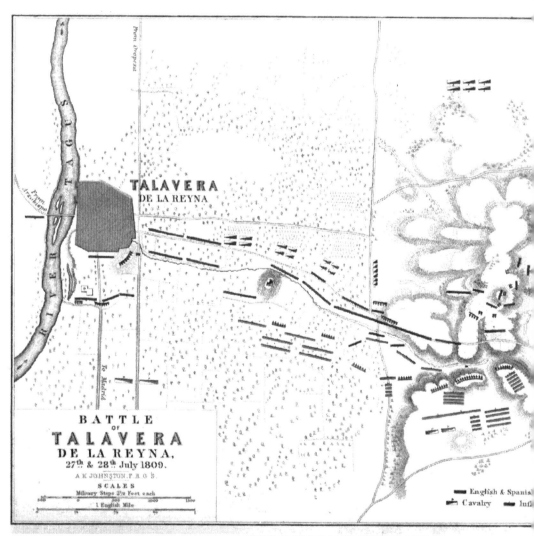

RIVER TAGUS

TALAVERA
DE LA REYNA

From Arzobispo

From Oropesa

To Madrid

BATTLE
OF
TALAVERA
DE LA REYNA,
27th & 28th July 1809.

A.K.JOHNSTON, F.R.G.S.

SCALES
Military Steps 2½ Feet each

1 English Mile

English & Spanish
Cavalry        Inf

WILLIAM BLACKWOOD & SONS, EDINBURGH & LONDON.

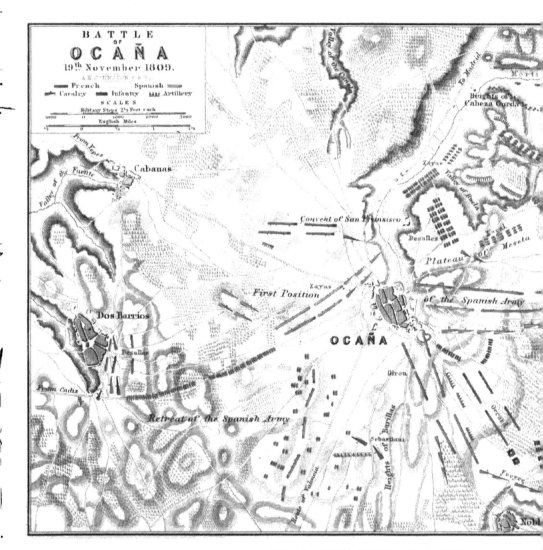

BATTLE
OF
OCAÑA
19th November 1809.
A.K. JOHNSTON F.R.S.

French —— Spanish ——
Cavalry —— Infantry —— Artillery

SCALES
Military Steps 2½ Feet each
English Miles

WILLIAM BLACKWOOD & SONS, EDINBURGH & LONDON.

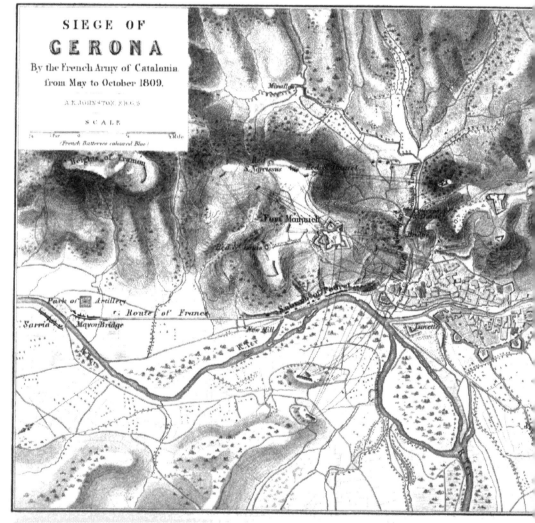

SIEGE OF
GERONA
By the French Army of Catalonia
from May to October 1809.

A.K.JOHNSTON, F.R.G.S.

SCALE

(French Batteries coloured Blue.)

WILLIAM BLACKWOOD & SONS. EDINBURGH & LONDON.

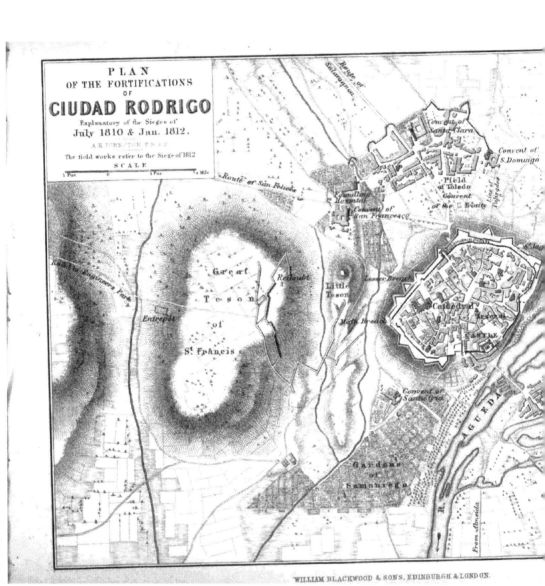

PLAN
OF THE FORTIFICATIONS
OF
CIUDAD RODRIGO
Explanatory of the Sieges of
July 1810 & Jan. 1812.
A K JOHNSTON F.R.G.S.
The field works refer to the Siege of 1812
SCALE

WILLIAM BLACKWOOD & SONS, EDINBURGH & LONDON.

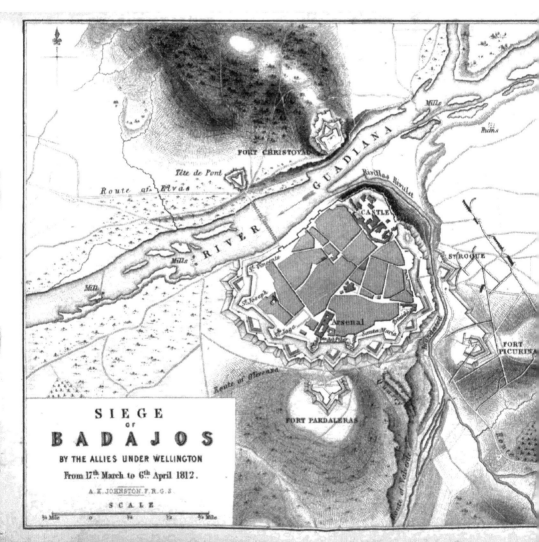

SIEGE
OF
BADAJOS
BY THE ALLIES UNDER WELLINGTON
From 17th March to 6th April 1812.

A.K. JOHNSTON, F.R.G.S.

SCALE

WILLIAM BLACKWOOD & SONS, EDINBURGH & LONDON.

PLAN OF THE SIEGE
OF
TARRAGONA

By the French Army of Aragon
4th May to 30th June 1811.

French    S C A L E    British

ROUTE of Valencia

Stone Bridge
Redoubt

TARRAGONA

Redoubt

Fort Royal
Bastion St John

Fort Francoli

LOWER TOWN

PORT

Pt of Milagro

Mole

ENGLISH SQUADRON
(Commodore Codrington)

M E D I T E R R A N E E

WILLIAM BLACKWOOD & SONS, EDINBURGH & LONDON

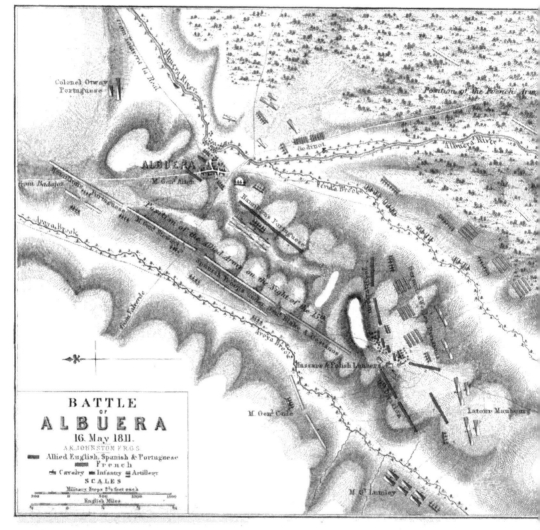

BATTLE
OF
ALBUERA
16. May 1811.
A.K. JOHNSTON F.R.G.S.

Allied English, Spanish & Portuguese
French
Cavalry  Infantry  Artillery

SCALES
Military Steps 2½ feet each

English Miles

WILLIAM BLACKWOOD & SONS, EDINBURGH & LONDON

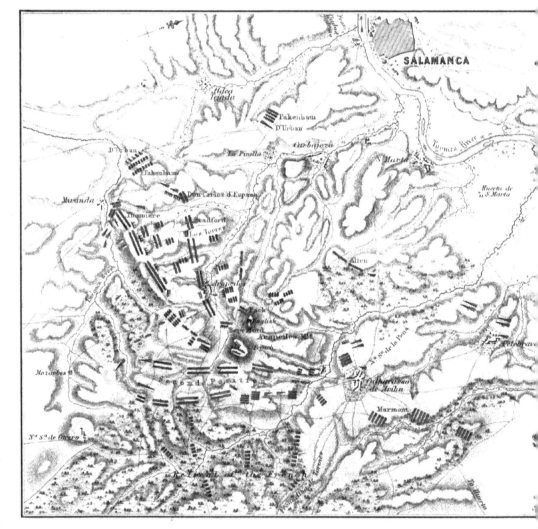

SALAMANCA

WILLIAM BLACKWOOD & SONS, EDINBURGH & LONDON.

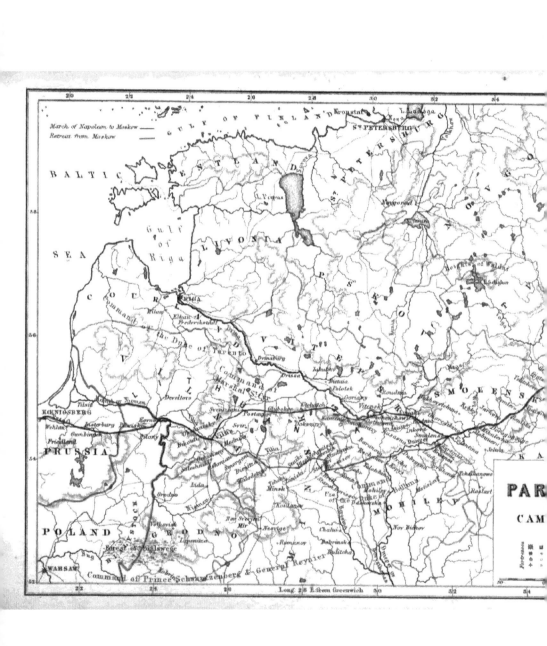

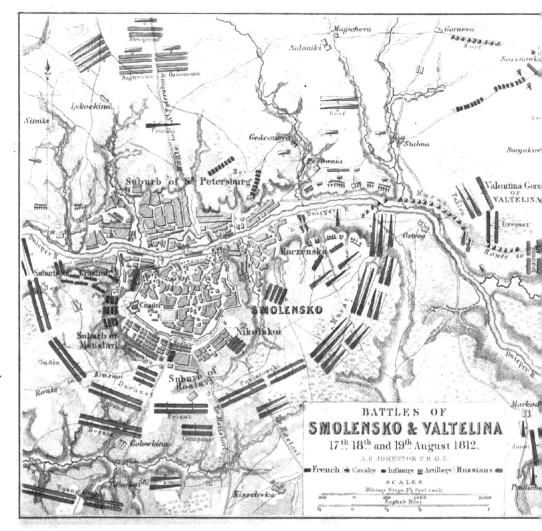

BATTLES OF
SMOLENSKO & VALTELINA
17th 18th and 19th August 1812.
A. K. JOHNSTON F.R.G.S.

French ⬤ Cavalry ⬤ Infantry ⬤ Artillery Russians ⬤

SCALES
Military Steps 2½ feet each
500    0    500    1000    2000
English Miles

WILLIAM BLACKWOOD & SONS, EDINBURGH & LONDON

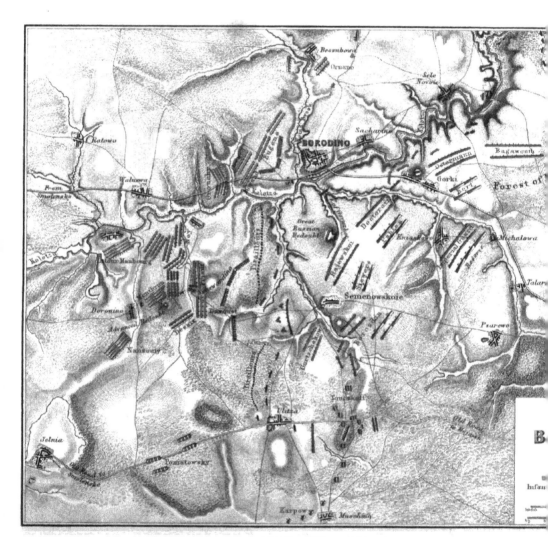

WILLIAM BLACKWOOD & SONS, EDINBURGH & LONDON

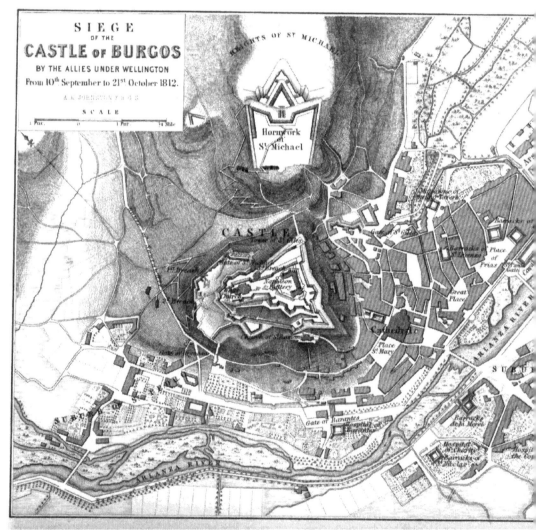

SIEGE
OF THE
CASTLE OF BURGOS
BY THE ALLIES UNDER WELLINGTON
From 10th September to 21st October 1812.

A.K. JOHNSTON F.R.G.S.

SCALE

HEIGHTS OF St MICHAEL

Hornwork
of
St Michael

CASTLE

ARLANZA RIVER

SUBURB OF SAN PEDRO

ARLANZA RIVER

WILLIAM BLACKWOOD & SONS EDINBURGH & LONDON

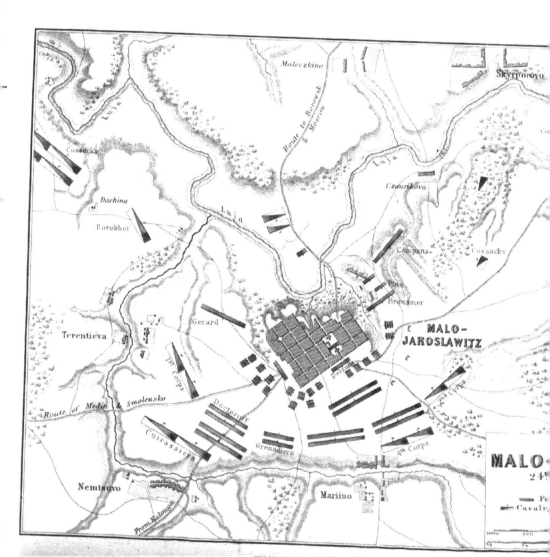

Maleczkino

Skyrporoyo

Luja

Cossacks

Luja

Czourikova

Dachina

Dorokhof

Luja

Compans

Cossacks

Guo

Broussier

Gerard

Terentieva

MALO-
JAROSLAWITZ

1er Corps

Deljon

5ᵗʰ Corps

Route of Medin & Smolensko

Doctoroff

Cuirassiers

Grenadiers

4ᵗʰ Corps

Nemtsovo

Marino

MALO-
24ᵗ

Fr
Cavalr

Prem Kalouga

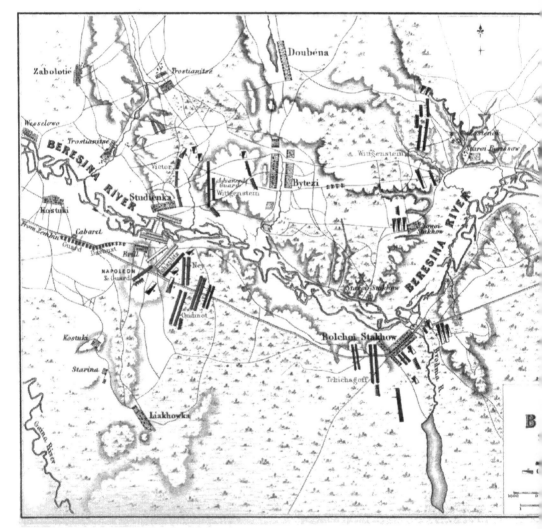

WILLIAM BLACKWOOD & SONS, EDINBURGH & LONDON.

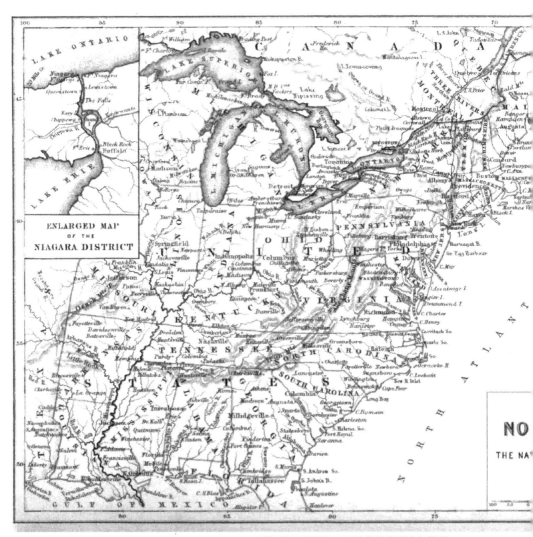

ENLARGED MAP
of the
NIAGARA DISTRICT

WILLIAM BLACKWOOD & SONS, EDINBURGH & LONDON.

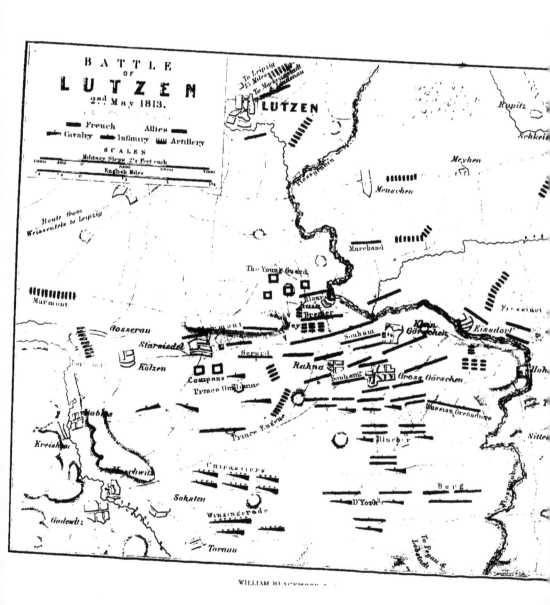

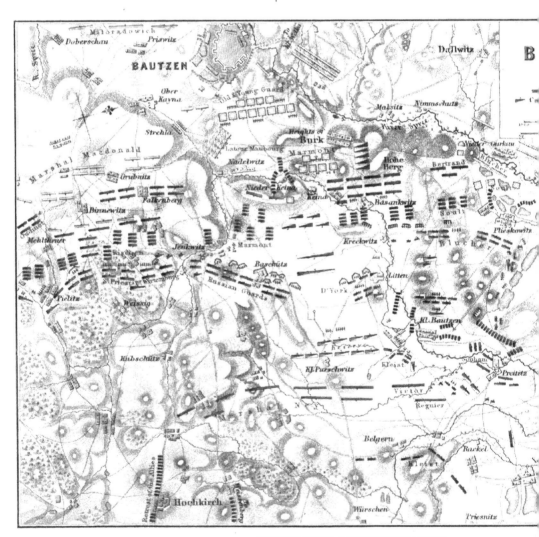

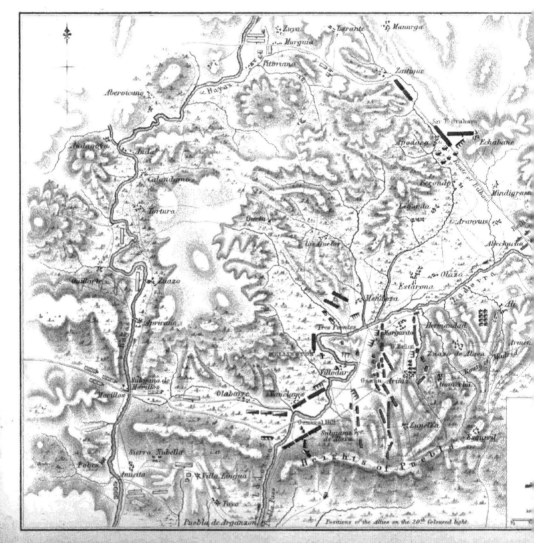

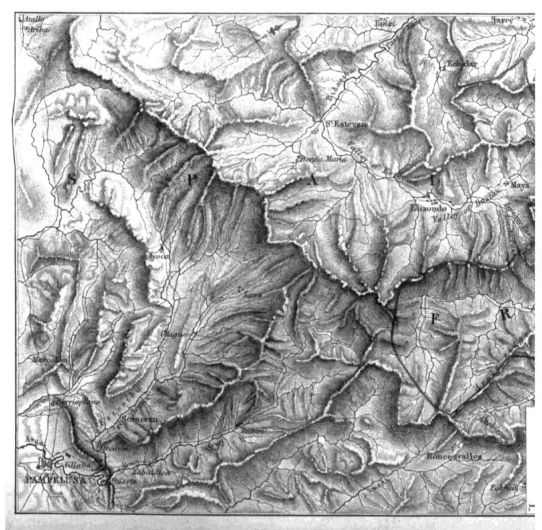

WILLIAM BLACKWOOD & SONS. EDINBURGH & LONDON.

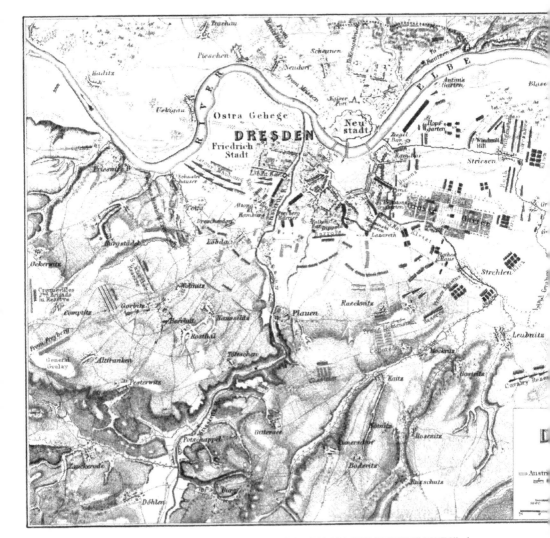

WILLIAM BLACKWOOD & SONS, EDINBURGH & LONDON.

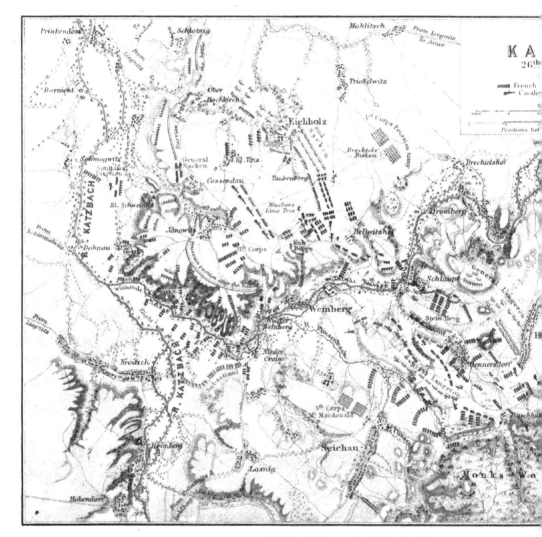

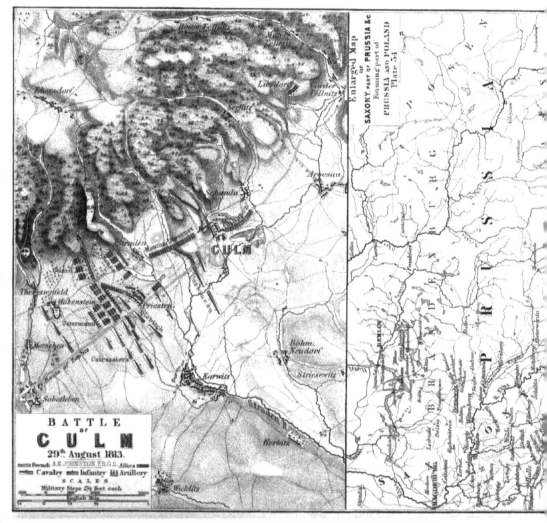

BATTLE
OF
CULM
29th August 1813.

French A.K.JOHNSTON, F.R.G.S. Allies
Cavalry   Infantry   Artillery
SCALES
Military Steps 2½ feet each

Enlarged Map
OF
SAXONY, PART OF PRUSSIA &c
Forming part of
PRUSSIA AND POLAND
Plate 54

WILLIAM BLACKWOOD & SONS, EDINBURGH & LONDON

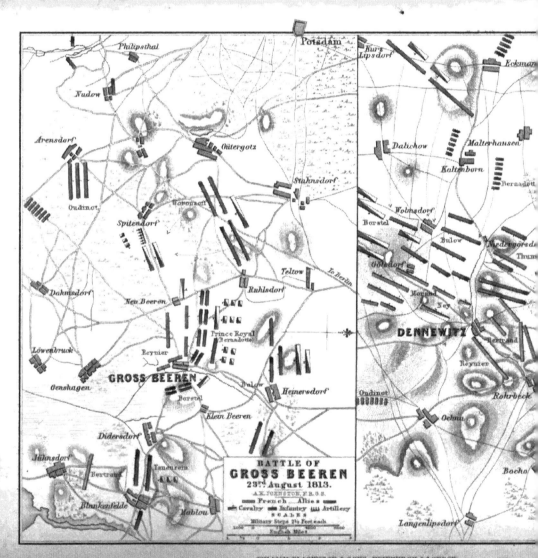

BATTLE OF
**GROSS BEEREN**
23RD August 1813.
A. K. JOHNSTON, F.R.S.E.
French ____ Allies ____
Cavalry ___ Infantry ___ Artillery
SCALES
Military Steps 2½ Feet each
English Miles

WILLIAM BLACKWOOD & SONS, EDINBURGH & LONDON.

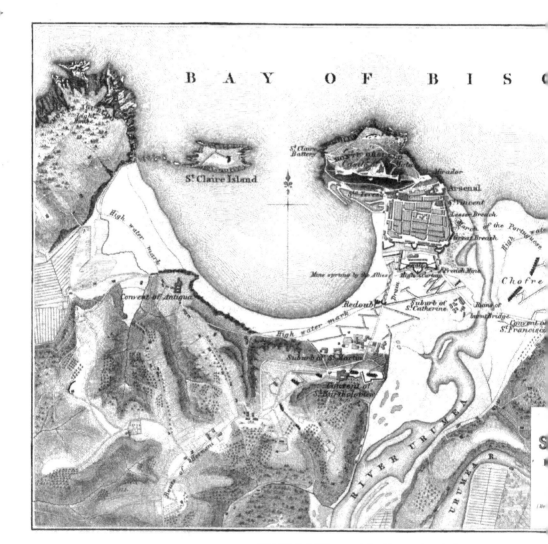

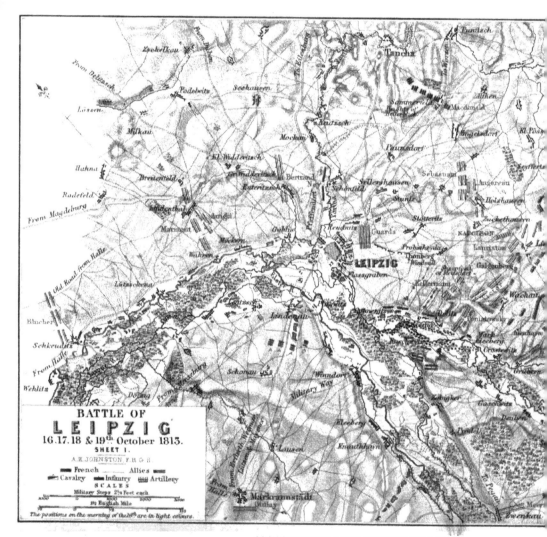

BATTLE OF
LEIPZIG
16, 17, 18 & 19th October 1813.
SHEET I.
A.K. JOHNSTON, F.R.G.S.

French ━━━  Allies ━━━
Cavalry  Infantry  Artillery

SCALES
Military Steps 2½ Feet each

½ English Mile

The positions on the morning of the 16th are in light colours.

WILLIAM BLACKWOOD & SONS, EDINBURGH & LONDON.

PLAN
of the
TOWN & SUBURBS
of
LEIPZIG

BATTLE OF
LEIPZIG
16th 18th & 19th October 1813.
SHEET 2.

A.K. JOHNSTON, F.R.G.S

French ........ Allies
Cavalry    Infantry    Artillery

SCALES

Military Steps 2½ Feet each

2½ English Mile

WILLIAM BLACKWOOD & SONS EDINBURGH & LONDON

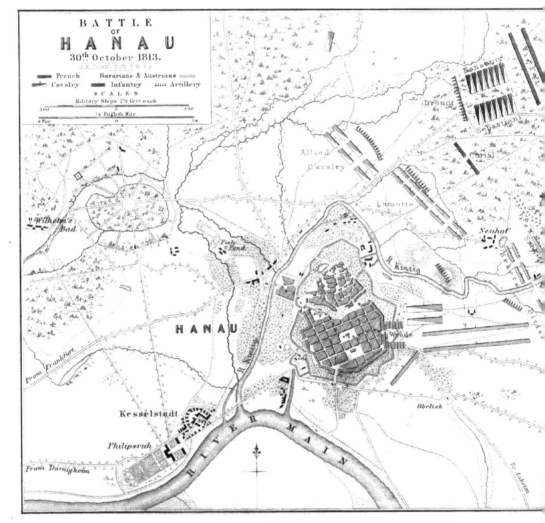

BATTLE
OF
HANAU
30th October 1813.
A.K.JOHNSTON F.R.G.S.

French        Bavarians & Austrians
Cavalry       Infantry        Artillery
SCALES
Military Steps 2½ feet each
500                ½ English Mile                500

Drouot

Allied
Cavalry

Christ

Lamotte

Wilhelm's
Bad

Neuhof

Fish Pond

R. Kinzig

HANAU

From Frankfort

R. Kinzig

Wyde

Kesselstadt

Obelisk

Philipsruh

From Dornigheim

R I V E R        M A I N

To Aschaffenburg

WILLIAM BLACKWOOD & SONS, EDINBURGH & LONDON

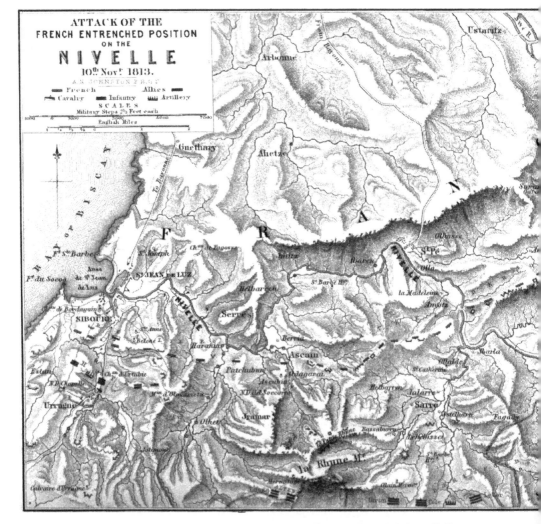

ATTACK OF THE
FRENCH ENTRENCHED POSITION
ON THE
# NIVELLE
10ᵗʰ Novʳ 1813.

A.K. JOHNSTON F.R.G.S

— French    — Allies
⊢ Cavalry    ▬ Infantry    ⁓ Artillery
SCALES
Military Steps 2½ Feet each
English Miles

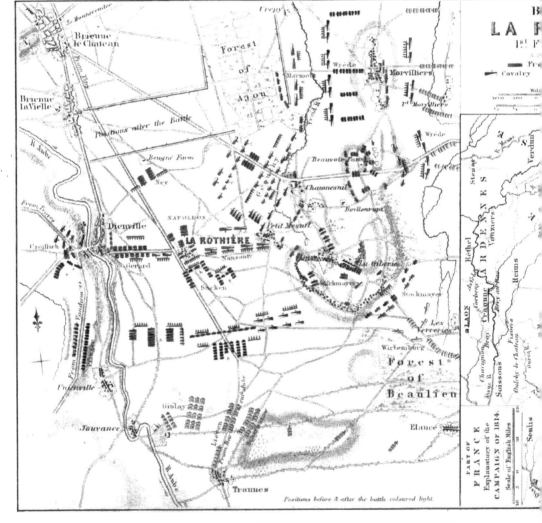

WILLIAM BLACKWOOD & SONS EDINBURGH & LONDON.

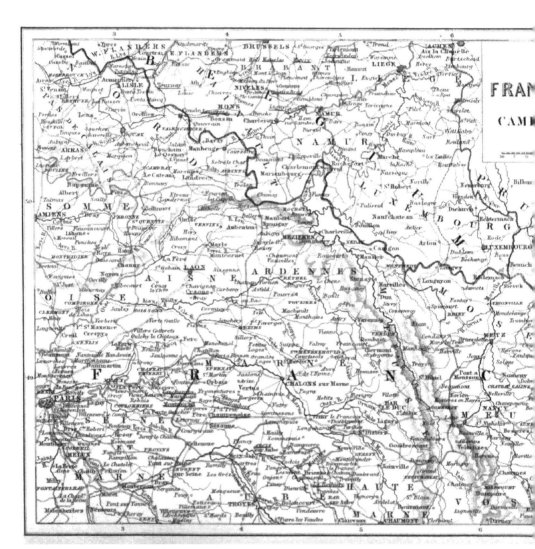

WILLIAM BLACKWOOD & SONS EDINBURGH & LONDON

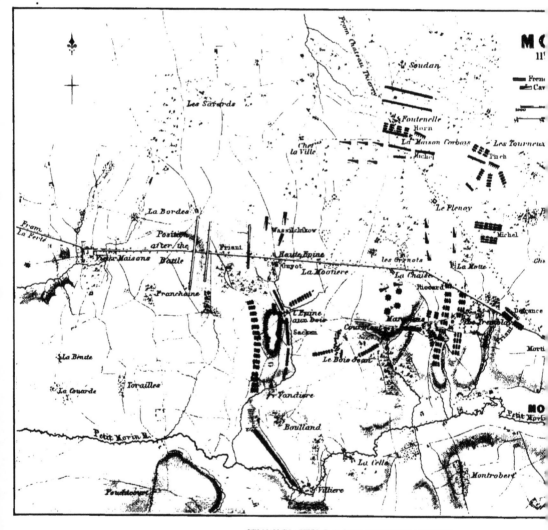

WILLIAM BLACKWOOD & SONS, EDINBURGH & LONDON

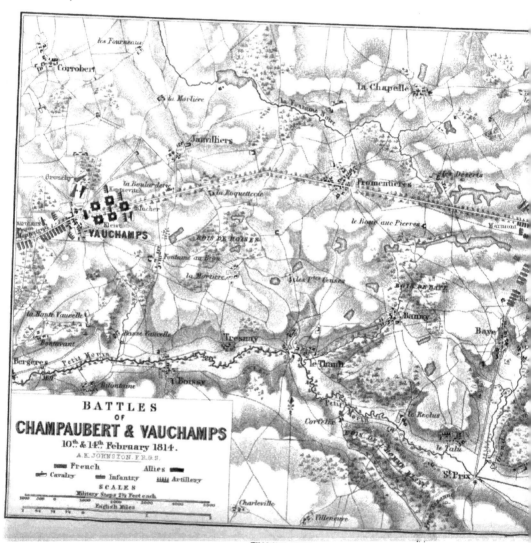

**BATTLES**
of
**CHAMPAUBERT & VAUCHAMPS**
10th & 14th February 1814.

A.K. JOHNSTON. F.R.G.S.

French      Allies

Cavalry    Infantry    Artillery

**SCALES**

Military Steps 2½ Feet each.

English Miles

WILLIAM BLACKWOOD & SONS, EDINBURGH & LONDON

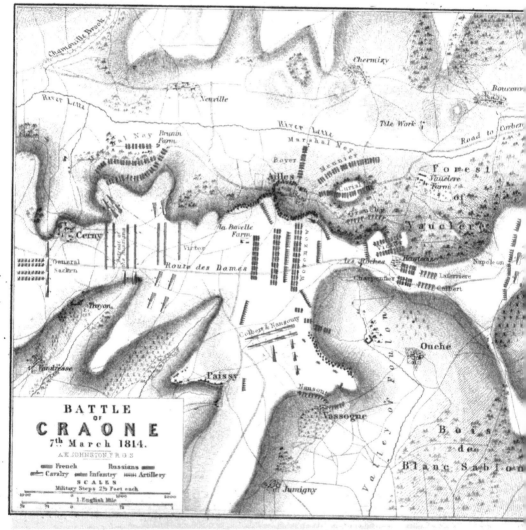

BATTLE
OF
CRAONE
7th March 1814.
A.K JOHNSTON, F.R.G.S

French        Russians
Cavalry   Infantry   Artillery
SCALES
Military Steps 2½ Feet each
1 English Mile

WILLIAM BLACKWOOD & SONS EDINBURGH & LONDON

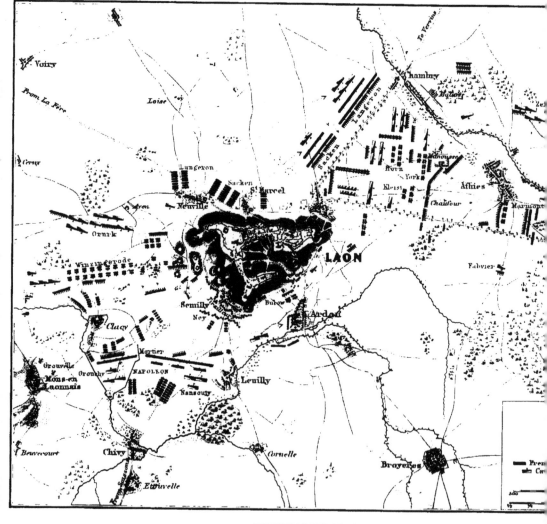

WILLIAM BLACKWOOD & SONS, EDINBURGH & LONDON

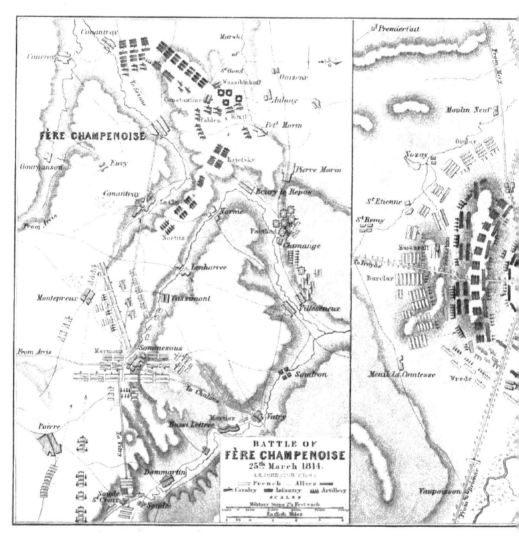

BATTLE OF
**FÈRE CHAMPENOISE**
25th March 1814.

A.K.JOHNSTON F.R.G.S

French ▨ Allies ■
Cavalry ▤ Infantry ▤ Artillery ▲
SCALE

Military Steps 2½ Feet each

English Miles

WILLIAM BLACKWOOD & SONS, EDINBURGH & LONDON.

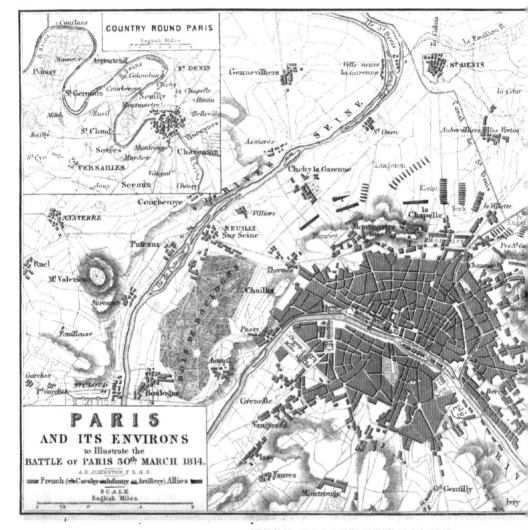

WILLIAM BLACKWOOD & SONS EDINBURGH & LONDON

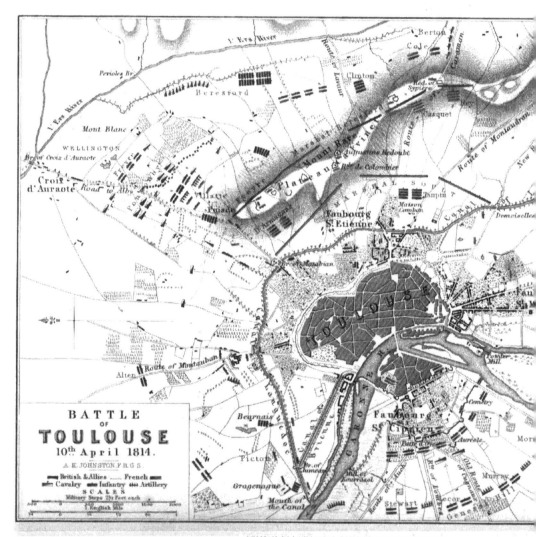

BATTLE
OF
TOULOUSE
10th April 1814.

A. K. JOHNSTON. F.R.G.S.

British & Allies ___ French
Cavalry ___ Infantry ___ Artillery

SCALES
Military Steps 2½ Feet each.

1 English Mile

WILLIAM BLACKWOOD & SONS, EDINBURGH & LONDON.

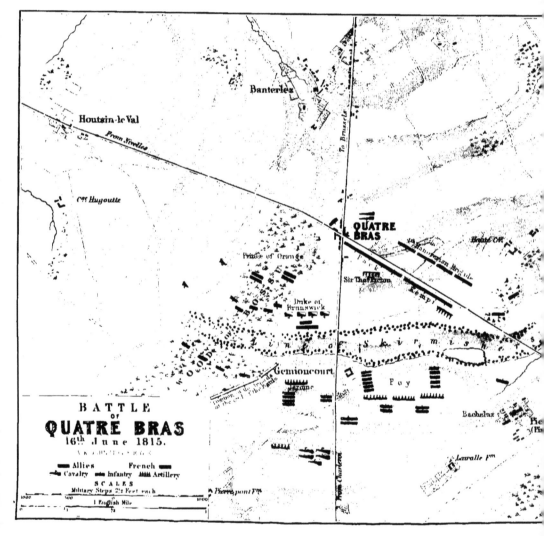

Banterlez

Houtain-le Val

From Nivelles

C.<sup>tt</sup> Huyoutte

QUATRE BRAS

Prince of Orange

Sir Tho.<sup>s</sup> Picton

Duke of Brunswick

Gemioncourt

Foy

Bachelaz

Lavalle F.<sup>m</sup>

BATTLE
OF
**QUATRE BRAS**
16.<sup>th</sup> June 1815.

Allies ▬   French ▬
Cavalry   Infantry   Artillery
S C A L E S
Military Steps 2½ Feet each
1 English Mile

Pierrepont F.<sup>m</sup>

WILLIAM BLACKWOOD & SONS EDINBURGH & LONDON

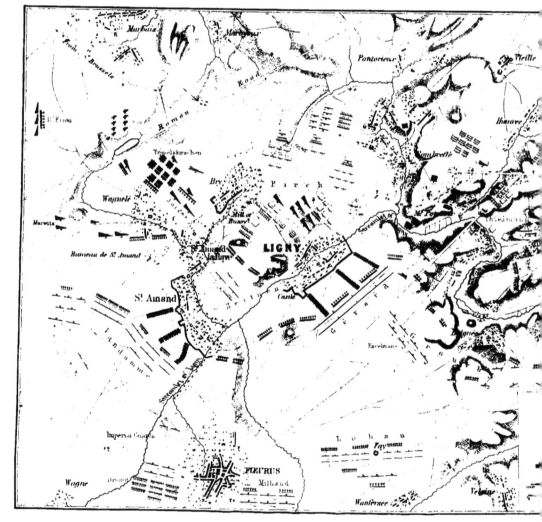

WILLIAM BLACKWOOD & SONS EDINBURGH & LONDON

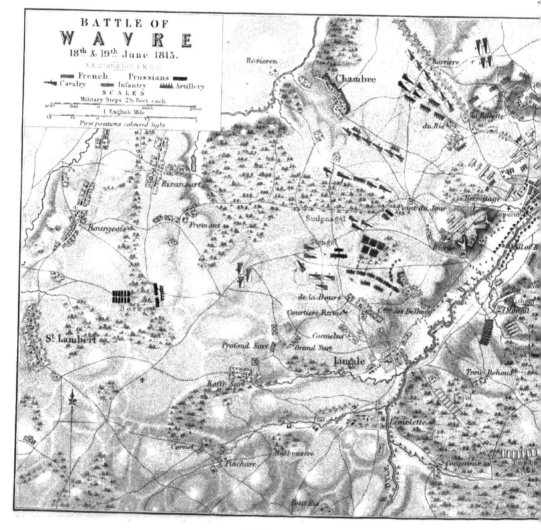

BATTLE OF
WAVRE
18th & 19th June 1815.

A K JOHNSTON F R G S

French     Prussians
Cavalry     Infantry     Artillery
SCALES
Military Steps 2½ Feet each

1 English Mile

First positions coloured light

WILLIAM BLACKWOOD & SONS EDINBURGH & LONDON.

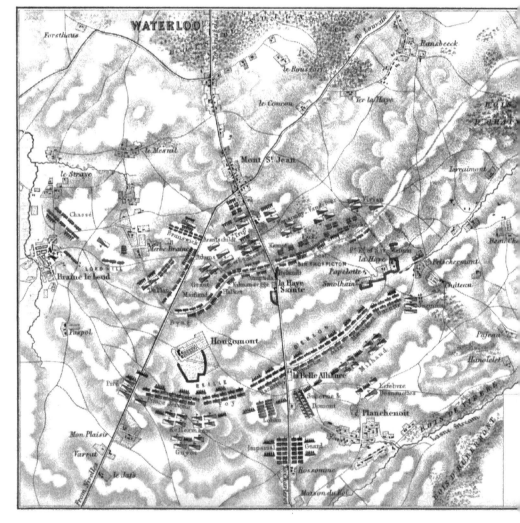

WILLIAM BLACKWOOD & SONS, EDINBURGH & LONDON

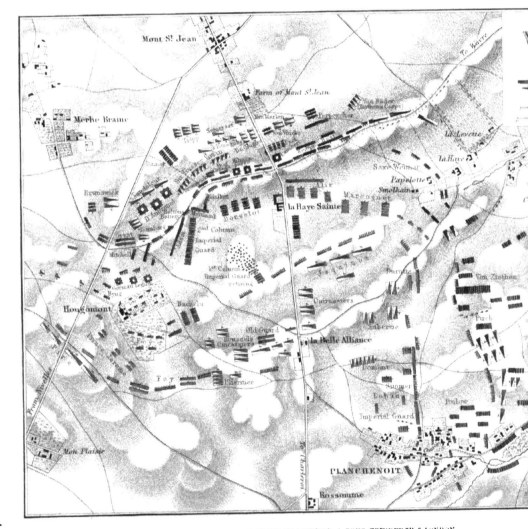

Mont St Jean

Merbe Braine

Farm of Mont St Jean

Brunswick

la Haye Sainte

Hougomont

la Belle Alliance

Foy

Mon Plaisir

PLANCHENOIT

Rossomme

WILLIAM BLACKWOOD & SONS, EDINBURGH & LONDON